Martin Steinbach

On the Stability of
Objective Structures

λογος

Augsburger Schriften zur Mathematik, Physik und Informatik
Band 38

Edited by:
Professor Dr. B. Schmidt
Professor Dr. B. Aulbach
Professor Dr. F. Pukelsheim
Professor Dr. W. Reif
Professor Dr. D. Vollhardt

All the code is available at the authors online repositories.

Bibliographic information published by the Deutsche Nationalbibliothek

The Deutsche Nationalbibliothek lists this publication in the
Deutsche Nationalbibliografie; detailed bibliographic data are
available in the Internet at http://dnb.d-nb.de .

ISBN 978-3-8325-4984-8
ISSN 1611-4256

Logos Verlag Berlin GmbH
Georg-Knorr-Str. 4, Geb. 10,
12681 Berlin
Tel.: +49 030 42 85 10 90
Fax: +49 030 42 85 10 92
INTERNET: http://www.logos-verlag.de

On the Stability of Objective Structures

Dissertation

zur Erlangung des akademischen Grades

Dr. rer. nat.

eingereicht an der

Mathematisch-Naturwissenschaftlich-Technischen Fakultät

der Universität Augsburg

von

Martin Steinbach

Augsburg, Dezember 2019

Erstgutachter: Prof. Dr. Bernd Schmidt (Universität Augsburg)

Zweitgutachter: Prof. Dr. Marco Hien (Universität Augsburg)

Drittgutachter: Prof. Dr. Richard D. James (University of Minnesota)

Mündliche Prüfung: 28. Oktober 2020

Contents

1. Introduction

In this thesis we consider a physical system made up of atoms which are in a static mechanical equilibrium. Thus, we have zero temperature, i. e., each atom has zero velocity, and the net force on each atom is zero. The atoms form an objective (atomic) structure like a lattice, graphene or a nanotube. Objective structures are defined by means of group theory which enables us to capitalize on their high symmetry. The atoms interact via a potential, e. g., the Lennard-Jones potential, which implicitly defines the configurational energy on the space of all periodic displacements. We consider only small displacements; particularly we are in the elasticity regime. The aim of this thesis is a theory of the (local) stability of the objective structure in this atomistic model. Usually, stability is defined by a second derivative test: An object at an equilibrium point is called stable if the second derivative of the configurational energy (at the equilibrium point) is coercive with respect to an appropriate seminorm. In this thesis we study which seminorm is appropriate for this stability condition. Further, we provide an efficient algorithm which checks the stability of an objective structure for a given interaction potential. We illustrate the algorithm by computing numerically the atomistic stability region of a toy model and a nanotube. In order to justify our choice of the seminorm, we also show that under certain reasonable assumptions, the second derivative of the configurational energy is bounded with respect to this seminorm. Thus, for a stable objective structure, the seminorm induced by the second derivative of the configurational energy is equivalent to our seminorm. Moreover, we show for a large class of objective structures as lattices and graphene that our second derivative test is indeed a sufficient condition for a local minimum of the configurational energy.

If the atoms form a lattice, the theory and the algorithm is well-understood, see, e. g., [40]. In this thesis we generalize the results from lattices to objective (atomic) structures, i. e., we assume that the set of positions of the atoms is equal to the orbit of a discrete subgroup of the Euclidean group under a point of the Euclidean space. We also assume that the stabilizer subgroup is trivial and thus we have a natural bijection between the discrete group and the atoms. The main issues for the generalization are the following:

(i) For a lattice there exists only one appropriate seminorm up to equivalence for the definition of stability. We show that for an objective structure there generally exist two appropriate nonequivalent seminorms: one for a stretched and one for a unstretched, i.e., stress-free, objective structure. For this purpose, we prove a discrete version of Korn's inequality and the equivalence of different seminorms for objective structures. If the objective structure is a lattice, this inequality implies the equivalence of the two appropriate seminorms.

(ii) For lattices it is well-known that their high symmetry implies that the second derivative of the configurational energy can be block diagonalized by means of Fourier analysis. We show that this generalizes to objective structures by means of harmonic analysis. The crux move is that due to the high symmetry of the objective structure which we have mathematically specified by the group, the operator associated with the second derivative of the configurational energy is left-translation-invariant. Thus, by harmonic analysis, the operator is a convolutional operator as well as a multiplier operator. Analogously, this is also true for the bilinear form which induces the appropriate seminorm. Roughly speaking, the Fourier transform simultaneously block diagonalizes the (infinite-dimensional) Hessian matrix and the matrix associated with the seminorm. This enables us to efficiently check the coerciveness of the second derivative of the configurational energy and hence the stability of an objective structure.

1.1. State of the art

The Cauchy-Born rule is a homogenization postulation to relate continuum theory to atomistic theory, see, e.g., the survey article [27]. If for a lattice the Cauchy-Born rule is valid, an elastic energy expression, more precisely a continuum energy functional with the linearized Cauchy-Born energy density, can be rigorously derived from an atomistic model as a Γ-limit. This was first done in one dimension [14] and then generalized to arbitrary dimensions [53, 16, 2, 12]. Also in plate theory, continuum models have been rigorously derived by Γ-convergence, see [32] for thick films and [51, 52] for thin films. For sheets, plates, and rods, continuum theories have also been derived with generalized Cauchy-Born rules; see, e.g., [60] for a general overview, [6] for the exponential Cauchy-Born rule that takes into account curvature, and [25] for the Saint-Venant's principle for nanotubes.

For a given interaction potential, in general it is hard to check the Cauchy-Born rule. Nevertheless, for a two-dimensional and for an arbitrary-dimensional mass-spring model, the validity and failure of the Cauchy-Born rule depending on its deformation has been proven in [33] and in [21], respectively. Also for graphene [31] and nanotubes [30], the validity of the Cauchy-Born rule has been rigorously proven.

There exist several definitions of stability which, in particular, provide a detailed analysis of the Cauchy-Born rule. The main difference between these stability definitions is the space of the allowed perturbations; see, e.g., [26]. For hard-loading devices which we consider in this thesis, periodic boundary conditions and periodic perturbations are an appropriate model, see [17]. As mentioned above, for lattices with periodic boundary conditions, the definition of stability in the atomistic model by Hudson and Ortner [40] is natural. Their definition requires that the second derivative is not only positive definite but also coercive. Moreover, they rigorously derive an algorithm such that they can numerically determine the stability region of a lattice. Based on this, in [17] the authors discuss the notion of stability in detail and derive the stability region and the failure of the Cauchy-Born rule analytically in an example. In [46] the authors generalize results of [40] to multi-lattices and they also discuss the equivalence class of the appropriate norm.

An application of this atomistic stability condition is that under its assumption, solutions of the equations of continuum elasticity with smooth body forces are asymptotically approximated by the corresponding atomistic equilibrium configurations. For both the static and the dynamic case, this has been proven for small displacements on a flat torus [24, 23], for the full space problem with a far-field condition [48], and for prescribed boundary values [17, 15].

In order to generalize the Cauchy-Born rule to a larger class of objects, James [41] defines objective structures by means of discrete subgroups of the Euclidean group. A characterization of the discrete subgroups of the three-dimensional and of an arbitrary-dimensional Euclidean group can be found in [18] and [3], respectively. In addition, the irreducible representations of space groups are well-known, see, e.g., [11, 13, 56]. To examine stability information of objective structures, James says that one should be able to do phonon analysis along the lines already done for crystal lattices. In [1] this is done for a three-dimensional objective structures which can be described by an abelian group.

1.2. Outline

In Chapter 2 we study discrete subgroups of the Euclidean group. Moreover, we define periodic functions on these subgroups and adapt well-
known theorems from harmonic analysis to our setting. In Section 2.1
we collect basic definitions and properties of the Euclidean group. In
Section 2.2 we state some well-known theorems about space groups. In
Section 2.3 we cite a characterization of discrete subgroups (of the Euclidean group). In order to define periodic boundary conditions, we after-
wards present a series of normal subgroups for a given discrete subgroup,
see Theorem 2.17. In Subsection 2.3.1 we collect some definitions and
theorems of harmonic analysis like the definition of the dual space and
the definition of induced representations. Up to a negligible set, the dual
space of a discrete subgroup is equal to a set of certain induced representations. In Subsection 2.3.2 we analyze these induced representations,
see Theorem 2.43. In Subsection 2.3.3 we define the inner product space
of all periodic functions. Then, we define the Fourier transform for both
periodic and absolutely summable functions and formulate well-known
theorems like the Plancherel formula for our setting. In Subsection 2.3.4
we generalize the Cauchy-Born rule to objective structures. Since we
are interested in the atomistic stability region, we also analyze the de-
pendence of the discrete group on the macroscopic deformation matrix.
In Subsection 2.3.5 we specify a series of normal subgroups for a given
discrete subgroup and represent the corresponding finite quotient groups
as semidirect products. The remainder of the thesis does not depend on
the results of this subsection. In Section 2.4 we describe an orbit of a
point under the action of the discrete subgroup by, for instance, its affine
dimension and a canonical coordinate system.

In Chapter 3 we define and examine the appropriate seminorms on
the space of all periodic displacements. The finite-dimensional kernel
of these seminorms corresponds to the isometries due to the invariance
of the configurational energy under isometries such as a translation. In
Section 3.1 we motivate the definition of the seminorm for the unstretched
case. In particular, we introduce and linearize our physical model; e.g.,
rotations are approximated by infinitesimal rotations. In Section 3.2 we
study this seminorm, prove its equivalence to similar seminorms and show
a discrete version of Korn's inequality. In the next section we define and
study the seminorm for the stretched case analogously. For the sake
of completeness, in Section 3.4 we consider a third seminorm which is
analogously defined to the two seminorms before. For a lattice, all of
these seminorms are equivalent, see Corollary 3.42. In Section 3.5 we

provide an example which shows that there exists no trivial formula of the seminorm in the Fourier space.

Chapter 4 is devoted to generalization the characterization of the stability constant of [40, Theorem 3.6(b)] from lattices to objective structures. This characterization resolves the central issue of the validation of the coerciveness and thus we have an algorithm to check stability. Moreover, we show that under realistic physical assumptions, the second derivative of the configurational energy is bounded by the seminorm. In Section 4.1 we define a very general many-body interaction potential with infinite range which we assume to be smooth and invariant under rotations. The interaction potential induces the configurational energy on the space of all periodic displacements. Moreover, we define stability in the atomistic model and a stability constant. In the next section we show how to check if an objective structure corresponds to a critical point of the configurational energy, see Corollary 4.16. For example, a simple lattice always corresponds to a critical point, see Corollary 4.17. In Section 4.3 we show for a large class of objective structures as lattices, that the stability of the objective structure is a sufficient condition that it corresponds to a local minimum of the configurational energy. In Section 4.4 we show that the second derivative of the configurational energy is bounded with respect to an appropriate seminorm under certain assumptions but particularly in dimension three, see Theorem 4.28, Theorem 4.34 and Theorem 4.39. In the next section we provide a characterization of the stability constant, see Theorem 4.51 and Theorem 4.54. In the proofs the Clifford theory is used. This theory describes the relation between representations of a group and of a normal subgroup. In Section 4.6 we summarize all results by providing an algorithm how to numerically check the stability of a given objective structure and of a interaction potential. Then we illustrate our results, first by means of a toy model and then by a nanotube. In particular, we see which seminorm is appropriate for the stretched and which seminorm is appropriate for the unstretched case.

1.3. Acknowledgments

I am very grateful to Prof. Bernd Schmidt for his supervision of my thesis, for always being supportive and for his guidance through each stage of the process. I would like to thank him for the interesting mathematical problem, for his confidence, and giving me so much freedom in my work.

I appreciate the pleasant working atmosphere I was able to experience at the Chair of Nonlinear Analysis and would like to thank all members for giving me a great time there. I enjoyed the mathematical and non-mathematical discussions, the coffee breaks and social activities like our yearly barbecue and visiting Scherneck. I am grateful to Veronika Auer-Volkmann for always cheering me up.

I also want to thank my friends, my siblings, particularly Simon Steinbach, and my parents for giving me support and always being there when I needed them. Finally, I want to thank my wife Elisa, who has been with me all these years and has made them the best years of my life.

2. Discrete subgroups of the Euclidean group

We will use the following notation. For all groups G and subsets $S_1, S_2 \subset G$ we denote

$$S_1 S_2 := \{s_1 s_2 \mid s_1 \in S_1, s_2 \in S_2\} \subset G$$

the product of group subsets. For all groups G, $S \subset G$, $n \in \mathbb{Z}$ and $g \in G$ we denote

$$S^n := \{s^n \mid s \in S\} \subset G$$

and

$$gS := \{gs \mid s \in S\} \subset G.$$

For two groups G, H we write $H < G$ if H is a proper subgroup of G and $H \lhd G$ if H is a normal subgroup of G. For a subset S of a group G we write $\langle S \rangle$ for the subgroup generated by S.

Moreover, let $\mathbb{N}$ be the set of all positive integers $\{1, 2, \dots\}$, $\mathbb{Z}_n$ be the group $\mathbb{Z}/(n\mathbb{Z})$, e_i be the ith standard coordinate vector $(0, \dots, 0, 1, \dots, 0)$ $\in \mathbb{R}^d$ and $I_n \in \mathbb{R}^{n \times n}$ be the identity matrix of size n. We use capital letters for matrices, and the direct sum of two matrices A and B is

$$A \oplus B := \begin{pmatrix} A & 0 \\ 0 & B \end{pmatrix}.$$

2.1. The Euclidean group

Let $d \in \mathbb{N}$ be the dimension. We denote the set of all Euclidean distance preserving transformations of $\mathbb{R}^d$ into itself by the *Euclidean group* $\mathrm{E}(d)$. The elements of $\mathrm{E}(d)$ are called *Euclidean isometries*. It is well-known that the Euclidean group $\mathrm{E}(d)$ can be described concretely as the outer semidirect product of $\mathbb{R}^d$ and $\mathrm{O}(d)$, the orthogonal group in dimension d:

$$\mathrm{E}(d) = \mathrm{O}(d) \ltimes \mathbb{R}^d.$$

The group operation is given by

$$(A_1, b_1)(A_2, b_2) = (A_1 A_2, b_1 + A_1 b_2)$$

for all $(A_1, b_1), (A_1, b_2) \in E(d)$, and the inverse of $(A, b) \in E(d)$ is

$$(A, b)^{-1} = (A^{-1}, -A^{-1}b).$$

Moreover, we define the homomorphism

$$\begin{aligned} L \colon E(d) &\to O(d) \\ (A, b) &\mapsto A \end{aligned}$$

and the map

$$\begin{aligned} \tau \colon E(d) &\to \mathbb{R}^d \\ (A, b) &\mapsto b. \end{aligned}$$

For all $(A, b) \in E(d)$ we call $L((A, b))$ the *linear component* and $\tau((A, b))$ the *translation component* of (A, b). Note that every isometry $g \in E(d)$ is uniquely defined by its linear and translation component:

$$g = (I_d, \tau(g))(L(g), 0).$$

We call an Euclidean isometry (A, b) a *translation* if $A = I_d$. All translations form the *group of translations* $\mathrm{Trans}(d)$, which is the abelian subgroup of $E(d)$ given by

$$\mathrm{Trans}(d) := \{I_d\} \ltimes \mathbb{R}^d.$$

We call a set of translations *linearly independent* if their translation components are linearly independent. The natural group action of $E(d)$ on $\mathbb{R}^d$ is given by

$$(A, b) \cdot x := Ax + b \qquad \text{for all } (A, b) \in E(d) \text{ and } x \in \mathbb{R}^d.$$

In this thesis we use a calligraphic font for subsets and particularly for subgroups of $E(d)$. For every group $\mathcal{G} < E(d)$ we denote the *orbit* of a point $x \in \mathbb{R}^d$ under the action of the group $\mathcal{G}$ by

$$\mathcal{G} \cdot x := \{g \cdot x \mid g \in \mathcal{G}\}.$$

We endow $E(d)$ with the subspace topology of the Euclidean space $\mathbb{R}^{d \times d} \times \mathbb{R}^d$ such that $E(d)$ is a topological group. It is well-known that a subgroup

$\mathcal{G} < \mathrm{E}(d)$ is discrete if and only if for every $x \in \mathbb{R}^d$ the orbit $\mathcal{G} \cdot x$ is discrete, see, e.g., [19, Exercise I.1.4]. In particular, every finite subgroup of $\mathrm{E}(d)$ is discrete.

A discrete group $\mathcal{G} < \mathrm{E}(d)$ is said to be *decomposable* if the group representation

$$\mathcal{G} \to \mathrm{GL}(d+1, \mathbb{C})$$

$$(A, b) \mapsto \begin{pmatrix} A & b \\ 0 & 1 \end{pmatrix}$$

is decomposable, i.e., there is a decomposition of $\mathbb{R}^{n+1}$ into the direct sum of two proper subspaces invariant under $\{ \begin{pmatrix} A & b \\ 0 & 1 \end{pmatrix} \mid (A, b) \in \mathcal{G} \}$. If this is not the case, the discrete group $\mathcal{G}$ is called *indecomposable*, see, e.g., [18, Appendix A.3]. An indecomposable discrete group $\mathcal{G} < \mathrm{E}(d)$ is also called a *(d-dimensional) space group*. In this thesis we will use the term space group. In section 2.2 and 2.3 we also present a (well-known) characterization of the space groups and the decomposable discrete subgroups of $\mathrm{E}(d)$, respectively, which does not use representation theory.

In the physically important case $d = 3$, all space groups and discrete decomposable subgroups of $\mathrm{E}(3)$ are well-known and classified, see, e.g., [5] and [47], respectively.

2.2. Space groups

The following theorem is well-known, see, e.g., [18, Appendix A.3].

Theorem 2.1. *Let $d \in \mathbb{N}$ be the dimension. A discrete subgroup of $\mathrm{E}(d)$ is a space group if and only if its subgroup of translations is generated by d linearly independent translations.*

Also the following theorem is well-known.

Theorem 2.2. *Let $\mathcal{G}$ be a d-dimensional space group and $\mathcal{T}$ its subgroup of translations. Then it holds:*

(i) The group $\mathcal{T}$ is a normal subgroup of $\mathcal{G}$ and isomorphic to $\mathbb{Z}^d$.

(ii) The point group $L(\mathcal{G})$ of $\mathcal{G}$ is finite.

(iii) The map

$$\mathcal{G}/\mathcal{T} \to L(\mathcal{G}), \quad (A, a)\mathcal{T} \mapsto A$$

is bijective and particularly, also $\mathcal{G}/\mathcal{T}$ is finite.

Proof. (i) This is clear by Theorem 2.1. (ii) See, e. g., [19, Theorem I.3.1]. (iii) It is easy to see that the map is bijective and by (ii) the set $\mathcal{G}/\mathcal{T}$ is finite. $\square$

Corollary 2.3. *Let $\mathcal{G}$ be a d-dimensional space group and $\mathcal{T}$ its subgroup of translations. Then for all $N \in \mathbb{N}$ the set $\mathcal{T}^N$ is a normal subgroup of $\mathcal{G}$ and isomorphic to $\mathbb{Z}^d$.*

Proof. This is clear by Theorem 2.2(i). $\square$

2.3. Discrete subgroups of the Euclidean group

Two subgroups $\mathcal{G}_1, \mathcal{G}_2 < \mathrm{E}(d)$ are termed *conjugate* subgroups under the group $\mathrm{E}(d)$ if there exists some $g \in \mathrm{E}(d)$ such that $g^{-1}\mathcal{G}_1 g = \mathcal{G}_2$. Note that every conjugation of a subgroup of $\mathrm{E}(d)$ under $\mathrm{E}(d)$ corresponds to a coordinate transformation in $\mathbb{R}^d$.

Now we characterize the discrete subgroups of $\mathrm{E}(d)$. For this purpose for all $d_1, d_2 \in \mathbb{N}$ we define the group homomorphism

$$\oplus \colon \mathrm{O}(d_1) \times \mathrm{E}(d_2) \to \mathrm{E}(d_1 + d_2)$$

$$(A_1, (A_2, b_2)) \mapsto A_1 \oplus (A_2, b_2) := \left(\begin{pmatrix} A_1 & 0 \\ 0 & A_2 \end{pmatrix}, \begin{pmatrix} 0 \\ b_2 \end{pmatrix} \right).$$

Theorem 2.4. *Let $d \in \mathbb{N}$ be the dimension and $\mathcal{G} < \mathrm{E}(d)$ be discrete. Then there exist $d_1, d_2 \in \mathbb{N}_0$ such that $d = d_1 + d_2$, a d_2-dimensional space group $\mathcal{S}$ and a discrete group $\mathcal{G}' < \mathrm{O}(d_1) \oplus \mathcal{S}$ such that $\mathcal{G}$ is conjugate under $\mathrm{E}(d)$ to $\mathcal{G}'$ and $\pi(\mathcal{G}') = \mathcal{S}$, where π is the natural surjective homomorphism $\mathrm{O}(d_1) \oplus \mathrm{E}(d_2) \to \mathrm{E}(d_2)$, $A \oplus g \mapsto g$.*

Proof. Let $d \in \mathbb{N}$ be the dimension and $\mathcal{G} < \mathrm{E}(d)$ be discrete. If $\mathcal{G}$ is a space group, the assertion is trivial. If $\mathcal{G}$ is finite, then $\mathcal{G}$ is conjugate under $\mathrm{E}(d)$ to a finite subgroup of $\mathrm{O}(d) \ltimes \{0_d\} \cong \mathrm{O}(d)$, see, e. g., [47, Section 4.12]. If $\mathcal{G}$ is an infinite decomposable discrete subgroup of $\mathrm{E}(d)$, the assertion is proven in [18, A.4 Theorem 2]. $\square$

Remark 2.5. Here $\mathrm{O}(d_1) \oplus \mathcal{S}$ is understood to be $\mathrm{O}(d)$ if $d_1 = d$ and to be $\mathcal{S}$ if $d_1 = 0$.

For the remainder of this section we fix the dimension $d \in \mathbb{N}$, the discrete group $\mathcal{G} < \mathrm{E}(d)$ and the quantities d_1, d_2, $\mathcal{T}$, $\mathcal{F}$, $\mathcal{S}$, $\mathcal{T}_{\mathcal{S}}$ by the following definition.

Definition 2.6. Let $d \in \mathbb{N}$ be the dimension. Let $d_1, d_2 \in \mathbb{N}_0$ be such that $d = d_1 + d_2$. Let $\mathcal{S}$ be a d_2-dimensional space group. Let $\mathcal{G} < \mathrm{O}(d_1) \oplus \mathcal{S}$ be discrete such that $\pi(\mathcal{G}) = \mathcal{S}$, where π is the natural surjective homomorphism $\mathrm{O}(d_1) \oplus \mathrm{E}(d_2) \to \mathrm{E}(d_2)$, $A \oplus g \mapsto g$. Let $\mathcal{F}$ be the kernel of $\pi|_{\mathcal{G}}$ and $\mathcal{T}_{\mathcal{S}}$ be the subgroup of translations of $\mathcal{S}$. Let $\mathcal{T} \subset \mathcal{G}$ such that the map $\mathcal{T} \to \mathcal{T}_{\mathcal{S}}$, $g \mapsto \pi(g)$ is bijective.

Remark 2.7.   (i) By Theorem 2.4 for every discrete group $\mathcal{G}' < \mathrm{E}(d)$ there exists some discrete group $\mathcal{G}$ as in Definition 2.6 such that $\mathcal{G}$ is conjugate to $\mathcal{G}'$ under $\mathrm{E}(d)$.

(ii) If $d_1 = 0$, we have $d_2 = d$, $\mathcal{G} = \mathcal{S}$, $\mathcal{T} = \mathcal{T}_{\mathcal{S}}$ and $\mathcal{F} = \{id\}$. If $d_1 = d$, we have $d_2 = 0$, $\mathcal{G}$ is finite, $\mathcal{G} = \mathcal{F}$ and $\mathcal{T} = \{id\}$.

(iii) The quantities d, d_1, d_2, $\mathcal{F}$, $\mathcal{S}$ and $\mathcal{T}_{\mathcal{S}}$ are uniquely defined by $\mathcal{G}$. In general for given $\mathcal{G}$ there is no canonical choice for $\mathcal{T}$, see Example 2.9.

(iv) Let $\mathcal{G}$ be given. In general, for every choice of $\mathcal{T}$ the set $\mathcal{T}$ is not a subset of $\mathrm{Trans}(d)$, see Example 2.8. Moreover, in general for every choice of $\mathcal{T}$ the set $\mathcal{T}$ is not a group and the elements of $\mathcal{T}$ do not commute, see Example 2.10.

(v) Let $\mathcal{G}$ be given. One possible choice for $\mathcal{T}$ is the following. Let $t_1, \ldots, t_{d_2} \in \mathcal{T}_{\mathcal{S}}$ be such that $\{t_1, \ldots, t_{d_2}\}$ generates $\mathcal{T}_{\mathcal{S}}$. For all $i \in \{1, \ldots, d_2\}$ let $g_i \in \mathcal{G}$ such that $\pi(g_i) = t_i$. Upon this, we define

$$\mathcal{T} = \{g_1^{n_1} \ldots g_{d_2}^{n_{d_2}} \mid n_1, \ldots, n_{d_2} \in \mathbb{Z}\}.$$

For the following example and the remainder of the thesis for all angles $\alpha \in \mathbb{R}$ we define the rotation matrix

$$R(\alpha) := \begin{pmatrix} \cos(\alpha) & -\sin(\alpha) \\ \sin(\alpha) & \cos(\alpha) \end{pmatrix} \in \mathrm{O}(2). \tag{2.1}$$

Example 2.8 (Helical groups). Let $d_1 = 2$, $d_2 = 1$, $\alpha \in \mathbb{R}$ be an angle, $n \in \mathbb{N}$,

$$\mathcal{T} = \Big\langle R(\alpha) \oplus (I_1, 1) \Big\rangle, \quad \mathcal{F} = \Big\langle R(2\pi/n) \oplus (I_1, 0) \Big\rangle$$

and

$$\mathcal{P} = \Big\langle \begin{pmatrix} 1 & 0 \\ 0 & -1 \end{pmatrix} \oplus (-I_1, 0) \Big\rangle.$$

Then $\mathcal{T}$ is isomorphic to $\mathbb{Z}$, $\mathcal{F}$ is a cyclic group of order n, $\mathcal{P}$ is a group of order 2 and $\mathcal{F}\mathcal{P}$ a dihedral group of order $2n$. Moreover, $\mathcal{T}$, $\mathcal{T}\mathcal{F}$,

$\mathcal{TP}$ and $\mathcal{TFP}$ are decomposable discrete subgroups of E(3). If we have $\alpha \in \mathbb{R} \setminus (2\pi\mathbb{Q})$, the groups $\mathcal{T}$, $\mathcal{TF}$, $\mathcal{TP}$ and $\mathcal{TFP}$ are so called *helical groups*, i. e. infinite discrete subgroups of the Euclidean group E(3) which do not contain any translation except the identity.

Example 2.9 (The choice of $\mathcal{T}$ is not unique.). Let $t = (I_1, 1)$, $\mathcal{F}_0 = \{I_2, R(\pi)\}$, $\mathcal{S} = \mathcal{T}_\mathcal{S} = \langle t \rangle$ and

$$\mathcal{G} = \left\{ (R(n\pi/2)F) \oplus t^n \,\middle|\, n \in \mathbb{Z}, F \in \mathcal{F}_0 \right\} < \mathrm{E}(3).$$

Then the choice $R(\pi/2) \oplus t \in \mathcal{T}$ as well as $R(3\pi/2) \oplus t \in \mathcal{T}$ is possible. In particular, the choice of $\mathcal{T}$ is not unique.

Example 2.10. We present a discrete group $\mathcal{G} < \mathrm{E}(8)$ such that for every choice of $\mathcal{T}$ the set $\mathcal{T}$ is not a group and the elements of $\mathcal{T}$ do not commute.
Let $\alpha_1, \alpha_2 \in \mathbb{R} \setminus (2\pi\mathbb{Q})$ be angles, $R_1 = R(\alpha_1)$, $R_2 = R(\alpha_2)$, $R_3 = R(\pi/2)$, $S = \left(\begin{smallmatrix} 1 & 0 \\ 0 & -1 \end{smallmatrix} \right)$, $t_1 = (I_2, e_1)$ and $t_2 = (I_2, e_2)$. Then we have $\langle R_1 \rangle \cong \mathbb{Z}$, $\langle R_2 \rangle \cong \mathbb{Z}$, and $\langle R_3, S \rangle < \mathrm{O}(2)$ is a dihedral group. Let $\mathcal{S} = \mathcal{T}_\mathcal{S} = \{ t_1^{n_1} t_2^{n_2} \,|\, n_1, n_2 \in \mathbb{Z} \}$,

$$\mathcal{G} := \left\{ \left(R_1^{n_1} \oplus R_2^{n_2} \oplus (S^{n_1} R_3^{n_2+m}) \right) \oplus \left(t_1^{n_1} t_2^{n_2} \right) \,\middle|\, n_1, n_2 \in \mathbb{Z}, m \in \{0, 2\} \right\}$$
$$22 < \mathrm{E}(8)$$

and $\pi \colon \mathcal{G} \to \mathcal{S}$ be the natural surjective homomorphism with kernel $\mathcal{F} = \{id, (I_4 \oplus R_3^2) \oplus id_{\mathrm{E}(2)}\}$. Let $\mathcal{T} \subset \mathcal{G}$ such that the map $\mathcal{T} \to \mathcal{T}_\mathcal{S}$, $g \mapsto \pi(g)$ is bijective. Since $t_1, t_2 \in \mathcal{T}_\mathcal{S}$, there exist $m_1, m_2 \in \{0, 2\}$ such that $t_1' := (R_1 \oplus I_2 \oplus (SR_3^{m_1})) \oplus t_1 \in \mathcal{T}$ and $t_2' := (I_2 \oplus R_2 \oplus R_3^{1+m_2}) \oplus t_2 \in \mathcal{T}$. We have $t_1' t_2' \neq t_2' t_1'$ since

$$t_1' t_2' (t_1')^{-1} (t_2')^{-1} = \left(I_4 \oplus (SR_3^{m_1} R_3^{1+m_2} R_3^{-m_1} SR_3^{-1-m_2}) \right) \oplus id_{\mathrm{E}(2)}$$
$$= (I_4 \oplus R_3^2) \oplus id_{\mathrm{E}(2)}. \tag{2.2}$$

Thus, the elements of $\mathcal{T}$ do not commute.
Now we suppose that $\mathcal{T}$ is a group. Since $\pi^{-1}(id_{\mathrm{E}(2)}) = \mathcal{F}$ and by (2.2), we have $\pi^{-1}(id_{\mathrm{E}(2)}) \subset \mathcal{T}$. This contradicts the claim that $\pi|_\mathcal{T}$ is bijective. Thus, $\mathcal{T}$ is not a group.

The following lemma characterizes the group $\mathcal{G}$.

Lemma 2.11. *(i) The group $\mathcal{F}$ is finite.*

(ii) For all $n \in \mathbb{N}$ the set $\mathcal{T}^n \mathcal{F}$ is independent of the choice of $\mathcal{T}$, and it holds
$$\mathcal{T}^n \mathcal{F} \lhd \mathcal{G}.$$
In particular, it holds $\mathcal{T}\mathcal{F} \lhd \mathcal{G}$.

(iii) The map $\mathcal{G}/\mathcal{T}\mathcal{F} \to \mathcal{S}/\mathcal{T}_\mathcal{S}$, $g\mathcal{T}\mathcal{F} \mapsto \pi(g)\mathcal{T}_\mathcal{S}$ is a group isomorphism, where $\pi \colon \mathcal{G} \to \mathcal{S}$ is the natural surjective homomorphism with kernel $\mathcal{F}$. In particular, $\mathcal{G}/\mathcal{T}\mathcal{F}$ is finite.

(iv) For all $n \in \mathbb{N}$ the map $\mathcal{T}_\mathcal{S} \to \mathcal{T}^n \mathcal{F}/\mathcal{F}$, $t \mapsto \varphi(t^n)\mathcal{F}$ is a group isomorphism, where $\varphi \colon \mathcal{T}_\mathcal{S}^n \to \mathcal{T}^n$ is the canonical bijection. In particular, the group $\mathcal{T}\mathcal{F}/\mathcal{F}$ is commutative.

(v) For all $n \in \mathbb{Z} \setminus \{0\}$ the map $\mathcal{T} \to \mathcal{T}^n$, $t \mapsto t^n$ is bijective.

Proof. Let $\pi \colon \mathcal{G} \to \mathcal{S}$ be the natural surjective homomorphism with kernel $\mathcal{F}$.

(i) Since $\mathcal{G}$ is discrete, the group $\mathcal{F}$ is discrete. Moreover, $\mathcal{F}$ is a subgroup of $\mathrm{O}(d_1) \oplus \{id_{\mathrm{E}(d_2)}\}$. Thus, the group $\mathcal{F}$ is finite.

(ii) Let $n \in \mathbb{N}$. The set $\mathcal{T}^n \mathcal{F}$ is the preimage of $\mathcal{T}_\mathcal{S}^n$ under π. Since $\mathcal{T}_\mathcal{S}^n$ is a normal subgroup of $\mathcal{S}$, the set $\mathcal{T}^n \mathcal{F}$ is a normal subgroup of $\mathcal{G}$.

(iii) This is clear, since $\mathcal{T}\mathcal{F}$ is the preimage of $\mathcal{T}_\mathcal{S}$ under π.

(iv) Let $n \in \mathbb{N}$. Since $\mathcal{T}_\mathcal{S}$ is isomorphic to $\mathbb{Z}^{d_2}$, the map $\varphi_1 \colon \mathcal{T}_\mathcal{S} \to \mathcal{T}_\mathcal{S}^n$, $t \mapsto t^n$ is a group isomorphism. Since $\mathcal{F}$ is the kernel of π and $\mathcal{T}^n \mathcal{F}$ the preimage of $\mathcal{T}_\mathcal{S}^n$ under π, the map $\varphi_2 \colon \mathcal{T}^n \mathcal{F}/\mathcal{F} \to \mathcal{T}_\mathcal{S}^n$, $g\mathcal{F} \mapsto \pi(g)$ is an isomorphism. This implies the assertion, i.e. the map $\varphi_2^{-1} \circ \varphi_1$ is an isomorphism.

(v) Let $n \in \mathbb{Z} \setminus \{0\}$. The map $\psi \colon \mathcal{T} \to \mathcal{T}^n$, $t \mapsto t^n$ is surjective. Since the map $\mathcal{T}_\mathcal{S} \to \mathcal{T}_\mathcal{S}^n$, $t \mapsto t^n$ is injective, the map ψ is injective and thus, bijective. $\qquad \square$

Lemma 2.12. Let $m \in \mathbb{Z} \setminus \{0\}$ such that $\mathcal{T}^m$ is a group. Then, the map
$$\mathcal{T}_\mathcal{S} \to \mathcal{T}^m$$
$$t \mapsto \varphi(t)^m$$
is a group isomorphism, where $\varphi \colon \mathcal{T}_\mathcal{S} \to \mathcal{T}$ is the canonical bijection. In particular, $\mathcal{T}^m$ is isomorphic to $\mathbb{Z}^{d_2}$.
Furthermore, for all $n \in \mathbb{Z}$ it holds
$$\mathcal{T}^{nm} \lhd \mathcal{T}^m.$$

Proof. Let $m \in \mathbb{Z} \setminus \{0\}$ such that $\mathcal{T}^m$ is a group. Let $\pi \colon \mathcal{T}\mathcal{F} \to \mathcal{T}_{\mathcal{S}}$ be the natural surjective homomorphism with kernel $\mathcal{F}$. Let φ be the inverse function of $\pi|_{\mathcal{T}}$, i.e. $\varphi \colon \mathcal{T}_{\mathcal{S}} \to \mathcal{T}$ is the canonical bijection. The map

$$\psi_1 \colon \mathcal{T}_{\mathcal{S}} \to \mathcal{T}\mathcal{F}/\mathcal{F}, \quad t \mapsto \varphi(t)\mathcal{F}$$

is an isomorphism. Since $\mathcal{T}\mathcal{F}/\mathcal{F}$ is isomorphic to $\mathbb{Z}^{d_2}$ and $(\mathcal{T}\mathcal{F}/\mathcal{F})^m = \mathcal{T}^m\mathcal{F}/\mathcal{F}$, the map

$$\psi_2 \colon \mathcal{T}\mathcal{F}/\mathcal{F} \to \mathcal{T}^m\mathcal{F}/\mathcal{F}, \quad t \mapsto t^m$$

is an isomorphism. Since $\mathcal{T}^m$ is a group, the map

$$\psi_3 \colon \mathcal{T}^m \to \mathcal{T}^m\mathcal{F}/\mathcal{F}, \quad g \mapsto g\mathcal{F}$$

is an isomorphism. The map

$$\mathcal{T}_{\mathcal{S}} \to \mathcal{T}^m, \quad t \mapsto \varphi(t)^m$$

is equal to $\psi_3^{-1} \circ \psi_2 \circ \psi_1$ and thus, an isomorphism.
Let $n \in \mathbb{Z}$. Since $\mathcal{T}^m$ is isomorphic to $\mathbb{Z}^{d_2}$, we have $\mathcal{T}^{mn} = (\mathcal{T}^m)^n \triangleleft \mathcal{T}^m$. $\qquad \square$

Definition 2.13. We define the set

$$M_0 := \{m \in \mathbb{N} \,|\, \mathcal{T}^m \text{ is a normal subgroup of } \mathcal{G}\}.$$

Remark 2.14. Let $N \in \mathbb{N}$. Then, the quotient group $\mathcal{G}/\mathcal{T}^N$ is well-defined if and only if $N \in M_0$.

Proposition 2.15. *For all $m \in M_0$ the group $\mathcal{T}^m$ is a subgroup of the center of $\mathcal{T}\mathcal{F}$.*

Proof. Let $m \in M_0$, $t \in \mathcal{T}$ and $g \in \mathcal{T}\mathcal{F}$. By Lemma 2.11(iv) there exists some $f \in \mathcal{F}$ such that
$$g t^m = t^m g f.$$
Since $m \in M_0$, it follows
$$f = g^{-1} t^{-m} g t^m \in \mathcal{T}^m.$$
Since $\mathcal{T}^m \cap \mathcal{F} = \{id\}$, we have $f = id$, i.e. g and t^m commute. $\qquad \square$

Lemma 2.16. *The set M_0 is not empty.*

Proof. Since $\mathcal{F}$ is a normal subgroup of $\mathcal{G}$, for all $g \in \mathcal{G}$ the map

$$\varphi_g \colon \mathcal{F} \to \mathcal{F}, \quad f \mapsto g^{-1} f g$$

is a group automorphism. Let n be the order of the automorphism group of $\mathcal{F}$. For all $g \in \mathcal{G}$ it holds $\varphi_g^n = id$. Thus for all $g \in \mathcal{G}$ and $f \in \mathcal{F}$ we have

$$g^n f = f g^n, \tag{2.3}$$

i. e. g^n and f commute.

Now we show that for all $g, h \in \mathcal{TF}$ the elements $g^{n|\mathcal{F}|}$ and h commute. Let $g, h \in \mathcal{TF}$. Since $\mathcal{TF}/\mathcal{F}$ is commutative, there exists some $f \in \mathcal{F}$ such that

$$h^{-1} g^n h = g^n f.$$

With (2.3) it follows

$$h^{-1} g^{n|\mathcal{F}|} h = (h^{-1} g^n h)^{|\mathcal{F}|} = (g^n f)^{|\mathcal{F}|} = g^{n|\mathcal{F}|} f^{|\mathcal{F}|} = g^{n|\mathcal{F}|}. \tag{2.4}$$

Now we show that $\mathcal{T}^{n|\mathcal{F}|^2}$ is a subgroup of $\mathcal{TF}$. Let $t, s \in \mathcal{T}$. We have to show that $t^{n|\mathcal{F}|^2} s^{-n|\mathcal{F}|^2} \in \mathcal{T}^{n|\mathcal{F}|^2}$. Let $r \in \mathcal{T}$ and $f \in \mathcal{F}$ such that $ts^{-1} = rf$. Since $\mathcal{TF}/\mathcal{F}$ is commutative, there exists some $e \in \mathcal{F}$ such that $t^{n|\mathcal{F}|} s^{-n|\mathcal{F}|} = r^{n|\mathcal{F}|} e$. By (2.4) and (2.3) we have

$$t^{n|\mathcal{F}|^2} s^{-n|\mathcal{F}|^2} = (t^{n|\mathcal{F}|} s^{-n|\mathcal{F}|})^{|\mathcal{F}|} = (r^{n|\mathcal{F}|} e)^{|\mathcal{F}|} = r^{n|\mathcal{F}|^2} e^{|\mathcal{F}|} = r^{n|\mathcal{F}|^2}$$
$$\in \mathcal{T}^{n|\mathcal{F}|^2}.$$

Now we show that $\mathcal{T}^{n|\mathcal{F}|^2}$ is a normal subgroup of $\mathcal{G}$. Let $g \in \mathcal{G}$ and $t \in \mathcal{T}$. We have to show that

$$g^{-1} t^{n|\mathcal{F}|^2} g \in \mathcal{T}^{n|\mathcal{F}|^2}.$$

Since $\mathcal{T}^n \mathcal{F}$ is a normal subgroup of $\mathcal{G}$, there exist some $s \in \mathcal{T}$ and $f \in \mathcal{F}$ such that

$$g^{-1} t^n g = s^n f.$$

By (2.3) we have

$$g^{-1} t^{n|\mathcal{F}|^2} g = (g^{-1} t^n g)^{|\mathcal{F}|^2} = (s^n f)^{|\mathcal{F}|^2} = s^{n|\mathcal{F}|^2} f^{|\mathcal{F}|^2} = s^{n|\mathcal{F}|^2} \in \mathcal{T}^{n|\mathcal{F}|^2}.$$
$$\qquad \qquad \square$$

Theorem 2.17. *There exists a unique $m_0 \in \mathbb{N}$ such that $M_0 = m_0 \mathbb{N}$.*

Proof. We define the set

$$\widetilde{M_0} := \{m \in \mathbb{Z} \mid \mathcal{T}^m \text{ is a normal subgroup of } \mathcal{G}\}.$$

First we show that $\widetilde{M_0}$ is a subgroup of the additive group of integers $\mathbb{Z}$. It is clear that $0 \in \widetilde{M_0}$. Let $n_1, n_2 \in \widetilde{M_0}$. We have to show that $n_1 - n_2 \in \widetilde{M_0}$. Let $\varphi \colon \mathcal{T}_\mathcal{S} \to \mathcal{T}$ be the canonical bijection. By Proposition 2.15 and Lemma 2.12, for all $t, s \in \mathcal{T}_\mathcal{S}$ it holds

$$\varphi(t)^{n_1-n_2} \varphi(s)^{-(n_1-n_2)} = \varphi(t)^{n_1} \varphi(s)^{-n_1} \varphi(t)^{-n_2} \varphi(s)^{n_2}$$
$$= \varphi(ts^{-1})^{n_1} \varphi(ts^{-1})^{-n_2}$$
$$= \varphi(ts^{-1})^{n_1-n_2} \in \mathcal{T}^{n_1-n_2},$$

and thus, $\mathcal{T}^{n_1-n_2}$ is a group. It remains to show that $\mathcal{T}^{n_1-n_2}$ is a normal subgroup of $\mathcal{G}$. Without loss of generality we assume that $n_1, n_2 \neq 0$, i.e. $n_1 n_2 \neq 0$. Let $g \in \mathcal{G}$ and $t \in \mathcal{T}$. Since $\mathcal{T}^{n_1}, \mathcal{T}^{n_2} \triangleleft \mathcal{G}$, there exist some $s_1, s_2 \in \mathcal{T}$ such that $gt^{n_1}g^{-1} = s_1^{n_1}$ and $gt^{n_2}g^{-1} = s_2^{n_2}$. Since $s_1^{n_1 n_2} = gt^{n_1 n_2}g^{-1} = s_2^{n_1 n_2}$ and the map $\mathcal{T} \to \mathcal{T}^{n_1 n_2}$, $r \mapsto r^{n_1 n_2}$ is bijective, it holds $s_1 = s_2$. Now we have

$$gt^{n_1-n_2}g^{-1} = (gt^{n_1}g^{-1})(gt^{n_2}g^{-1})^{-1} = s_1^{n_1-n_2} \in \mathcal{T}^{n_1-n_2}.$$

By Lemma 2.16 and since $M_0 \subset \widetilde{M_0}$, the group $\widetilde{M_0}$ is nontrivial. Since every nontrivial subgroup of $\mathbb{Z}$ is equal to $n\mathbb{Z}$ for some $n \in \mathbb{N}$, see, e.g., [20, Article 36], there exists a unique $m_0 \in \mathbb{N}$ such that $\widetilde{M_0} = m_0\mathbb{Z}$. Now, we have

$$M_0 = \widetilde{M_0} \cap \mathbb{N} = m_0\mathbb{N}. \qquad \square$$

Remark 2.18.   (i) The proof of Lemma 2.16 shows that m_0 divides $|\mathcal{F}|^2 |\mathrm{Aut}(\mathcal{F})|$, where $m_0 \in \mathbb{N}$ is such that $M_0 = m_0\mathbb{N}$ and $\mathrm{Aut}(\mathcal{F})$ is the automorphism group of $\mathcal{F}$. In particular, we have an upper bound for m_0.

(ii) The group $\mathcal{G}$ is virtually abelian since for all $m \in M_0$ the index of the abelian subgroup $\mathcal{T}^m$ in $\mathcal{G}$ is $m^{d_2} |\mathcal{F}| |\mathcal{G}/\mathcal{T}\mathcal{F}|$ and thus, finite.

2.3.1. The dual space and induced representations

In this subsection we define some terms of representation theory. In our set-up it is not restrictive to only consider finite-dimensional representations, see Remark 2.20 below.

Definition 2.19. Let $\mathcal{H}$ be a finite group or a discrete subgroup of $E(d)$. A *representation* of $\mathcal{H}$ is a homomorphism $\rho\colon \mathcal{H} \to U(d_\rho)$, where $d_\rho \in \mathbb{N}$ is the *dimension* of ρ and $U(d_\rho)$ is the group of all unitary matrices in $\mathbb{C}^{d_\rho \times d_\rho}$. Two representations ρ, ρ' of $\mathcal{H}$ are said to be *equivalent* if $d_\rho = d_{\rho'}$ and there exists some $T \in U(d_\rho)$ such that

$$T^{\mathsf{H}}\rho(g)T = \rho'(g) \qquad \text{for all } g \in \mathcal{H}.$$

A representation ρ of $\mathcal{H}$ is said to be *irreducible* if the only subspaces of $\mathbb{C}^{d_\rho}$ invariant under $\{\rho(g) \mid g \in \mathcal{H}\}$ are $\{0\}$ and $\mathbb{C}^{d_\rho}$. Let $\widehat{\mathcal{H}}$ denote the set of all equivalence classes of irreducible representations of $\mathcal{H}$. One calls $\widehat{\mathcal{H}}$ the *dual space* of $\mathcal{H}$. If $\mathcal{N}$ is a normal subgroup of $\mathcal{H}$, then the group $\mathcal{H}$ acts on the set of all representations of $\mathcal{N}$ by

$$g \cdot \rho(n) := \rho(g^{-1}ng) \quad \text{for all } g \in \mathcal{H}, \text{ representations } \rho \text{ of } \mathcal{N} \text{ and } n \in \mathcal{N}.$$

For given representations $\rho_1, \ldots, \rho_n$ of $\mathcal{H}$, we define the *direct sum*

$$\oplus_{i=1}^{n}\rho_i\colon \mathcal{H} \to U(m)$$
$$g \mapsto \oplus_{i=1}^{n}(\rho_i(g)),$$

where $m = \sum_{i=1}^{n} d_{\rho_i}$. In a canonical way, the above group action and terms *dimension*, *irreducible* and *direct sum* are also defined for equivalence classes of representations.

Remark 2.20. In [45] the following theorem is proved for any locally compact group: There exists an integer $M \in \mathbb{N}$ such that the dimension of every irreducible representation is less than or equal to M if and only if there is an open abelian subgroup of finite index. This, in particular, applies to finite groups and discrete subgroups of $E(d)$.

A caveat on notation: For a representation and for an equivalence class of representations we use the symbol χ if it is one-dimensional and ρ otherwise. For every one-dimensional representation χ its equivalence class is a singleton which we also call a *representation* and denote χ. The following lemma is well-known.

Lemma 2.21. *Let $\chi, \rho, \rho_1, \rho_2$ be representations of a discrete group $\mathcal{H} <$ $E(d)$ such that χ is one-dimensional. Then it holds:*

(i) The map $\chi\rho$ is also a representation of the group $\mathcal{H}$.

(ii) If ρ is irreducible, then also $\chi\rho$ is irreducible.

(iii) If ρ_1 and ρ_2 are equivalent, then also $\chi\rho_1$ and $\chi\rho_2$ are equivalent.

Lemma 2.22. *Let $\mathcal{H} < \mathrm{E}(d)$ be discrete. Then we have*

$$g \cdot \rho = \rho \qquad \text{for all } g \in \mathcal{H} \text{ and } \rho \in \widehat{\mathcal{H}}.$$

Proof. This is well-known, see, e. g., [28, Subsection XII.1.3], but for the reader's convenience we give a proof. Let $\mathcal{H} < \mathrm{E}(d)$ be discrete, $g \in \mathcal{H}$, $\rho \in \widehat{H}$ and $\tilde{\rho}$ be a representative of ρ. Then we have

$$g \cdot \tilde{\rho}(h) = \tilde{\rho}(g^{-1}hg) = \tilde{\rho}(g)^{-1}\tilde{\rho}(h)\tilde{\rho}(g) \qquad \text{for all } h \in \mathcal{H}.$$

Thus the representations $g \cdot \tilde{\rho}$ and $\tilde{\rho}$ are equivalent and we have $g \cdot \rho = \rho$. $\qquad\square$

We define the induced representation as in [55, Section 8.2], where it is defined for finite groups.

Definition 2.23. Let $\mathcal{H} < \mathrm{E}(d)$ be discrete and $\mathcal{K}$ be a subgroup of $\mathcal{H}$ such that the index $n = |\mathcal{H} : \mathcal{K}|$ if finite. Choose a complete set of representatives $\{h_1, \ldots, h_n\}$ of the left cosets of $\mathcal{K}$ in $\mathcal{H}$. Suppose $\rho \colon \mathcal{K} \to \mathrm{U}(d_\rho)$ is a representation of $\mathcal{K}$. Let us introduce a dot notation in this context by setting

$$\dot{\rho}(g) := \begin{cases} \rho(g) & \text{if } g \in \mathcal{K} \\ 0_{d_\rho, d_\rho} & \text{else} \end{cases}$$

for all $g \in \mathcal{H}$. The *induced representation* $\mathrm{Ind}_{\mathcal{K}}^{\mathcal{H}} \rho \colon \mathcal{H} \to \mathrm{U}(nd_\rho)$ is defined by

$$\mathrm{Ind}_{\mathcal{K}}^{\mathcal{H}} \rho(g) = \begin{bmatrix} \dot{\rho}(h_1^{-1}gh_1) & \cdots & \dot{\rho}(h_1^{-1}gh_n) \\ \vdots & \ddots & \vdots \\ \dot{\rho}(h_n^{-1}gh_1) & \cdots & \dot{\rho}(h_n^{-1}gh_n) \end{bmatrix} \qquad \text{for all } g \in \mathcal{H}.$$

The *induced representation* of an equivalence class of representations is the equivalence class of the induced representation of a representative. Moreover, let $\mathrm{Ind}_{\mathcal{K}}^{\mathcal{H}}(\widehat{\mathcal{K}})$ denote the set of all induced representations of $\widehat{\mathcal{K}}$. We also write Ind instead of $\mathrm{Ind}_{\mathcal{K}}^{\mathcal{H}}$ if $\mathcal{K}$ and $\mathcal{H}$ are clear by context.

Remark 2.24. For a general locally compact group the definition of the induced representation is more complicated, see, e. g., [43, Chapter 2].

The following proposition is standard in Clifford theory.

Proposition 2.25. *Let $\mathcal{H} < \mathrm{E}(d)$ be discrete and $\mathcal{N}$ be a normal subgroup of $\mathcal{H}$ such that the index $|\mathcal{H} : \mathcal{N}|$ is finite. Then the map*

$$\widehat{\mathcal{N}}/\mathcal{H} \to \mathrm{Ind}_{\mathcal{N}}^{\mathcal{H}}(\widehat{\mathcal{N}})$$

$$\mathcal{H} \cdot \rho \mapsto \mathrm{Ind}_{\mathcal{N}}^{\mathcal{H}} \rho$$

is bijective, where $\widehat{\mathcal{N}}/\mathcal{H} = \{\mathcal{H} \cdot \rho \mid \rho \in \widehat{\mathcal{N}}\}$.

Proof. Let $\mathcal{H} < \mathrm{E}(d)$ be discrete and $\mathcal{N}$ be a normal subgroup of $\mathcal{H}$ such that the index $n = |\mathcal{H} : \mathcal{N}|$ is finite. Let $\{h_1, \ldots, h_n\}$ be a complete set of representatives of the cosets of $\mathcal{N}$ in $\mathcal{H}$ and φ be the map $\widehat{\mathcal{N}}/\mathcal{H} \to \mathrm{Ind}(\widehat{\mathcal{N}})$, $\mathcal{H} \cdot \rho \mapsto \mathrm{Ind}\,\rho$.

First we show that φ is well-defined. Let $\rho \in \widehat{\mathcal{N}}$ and $g \in \mathcal{H}$. Let σ be the permutation of $\{1, \ldots, n\}$ and $k_1, \ldots, k_n \in \mathcal{N}$ such that $gh_i = h_{\sigma(i)}k_{\sigma(i)}$ for all $i \in \{1, \ldots, n\}$. For all $h \in \mathcal{N}$ we have

$$\mathrm{Ind}(g \cdot \rho)(h) = U^{\mathsf{H}}(\mathrm{Ind}\,\rho(h))U$$

with

$$U = (\rho(k_1) \oplus \cdots \oplus \rho(k_n))(P_\sigma^{\mathsf{T}} \otimes I_{d_\rho}) \in \mathrm{U}(nd_\rho),$$

where P_σ is the permutation matrix $(\delta_{\sigma(i),j})_{ij}$.

It is clear that φ is surjective.

Now we show that φ is injective. Let $\rho, \rho' \in \widehat{\mathcal{N}}$ such that $\mathrm{Ind}\,\rho = \mathrm{Ind}\,\rho'$. Let $\tilde{\rho}$ and $\tilde{\rho}'$ be representatives of ρ and ρ', respectively. Since $\mathcal{N}$ is a normal subgroup, for all $g \in \mathcal{N}$ and $i, j \in \{1, \ldots, n\}$ we have $h_i^{-1}gh_j \in \mathcal{N}$ if and only if $i = j$. Thus we have

$$(\mathrm{Ind}\,\tilde{\rho})|_{\mathcal{N}} = \oplus_{i=1}^n h_i \cdot \tilde{\rho} \qquad \text{and} \qquad (\mathrm{Ind}\,\tilde{\rho}')|_{\mathcal{N}} = \oplus_{i=1}^n h_i \cdot \tilde{\rho}'. \qquad (2.5)$$

Since the representations $h_1 \cdot \rho, \ldots, h_n \cdot \rho$ and $h_1 \cdot \rho', \ldots, h_n \cdot \rho'$ are irreducible, by (2.5) there exists some $i \in \{1, \ldots, n\}$ such that $h_1 \cdot \rho = h_i \cdot \rho'$. Thus we have $\mathcal{H} \cdot \rho = \mathcal{H} \cdot \rho'$. $\qquad \square$

2.3.2. The induced representations $\mathrm{Ind}(\widehat{\mathcal{TF}})$

The following definition and Lemma 2.27 can be found in [44, Chapter 1].

Definition 2.26. A set $L \subset \mathbb{R}^n$ is a *lattice* if L is a subgroup of the additive group $\mathbb{R}^n$ which is isomorphic to the additive group $\mathbb{Z}^n$, and which spans the real vector space $\mathbb{R}^n$.

The *dual lattice* L^* (also called the *reciprocal lattice*) of a lattice $L \subset \mathbb{R}^n$ is the set

$$\{x \in \mathbb{R}^n \mid \langle x, y \rangle \in \mathbb{Z} \text{ for all } y \in L\}.$$

Lemma 2.27. *For every lattice in $\mathbb{R}^n$ its dual lattice is also a lattice.*

Proof. This is well-known, see, e.g., [44, Section 1.2]. For the reader's convenience we give a proof. Let L be a lattice and L^* its dual lattice. There exist $b_1, \ldots, b_n \in \mathbb{R}^n$ such that $\{b_1, \ldots, b_n\}$ generates L and is a basis of $\mathbb{R}^{d_2}$. For all $i \in \{1, \ldots, n\}$ there exists a unique $b_i' \in \mathbb{R}^n$ such that

$$\langle b_i', b_j \rangle = \delta_{ij} \qquad \text{for all } j \in \{1, \ldots, n\}.$$

It is easy to see that $\{b_1', \ldots, b_n'\}$ is a basis of $\mathbb{R}^n$ (called the dual basis of $\{b_1, \ldots, b_n\}$) and

$$L^* = \left\{ \sum_{i=1}^{n} m_i b_i' \,\middle|\, m_1, \ldots, m_n \in \mathbb{Z} \right\}. \qquad \square$$

Definition 2.28. We define the lattice

$$L_{\mathcal{S}} := \tau(\mathcal{T}_{\mathcal{S}}) < \mathbb{R}^{d_2}$$

and denote its dual lattice by $L_{\mathcal{S}}^*$.

Definition 2.29. For all $k \in \mathbb{R}^{d_2}$ we define the one-dimensional representation $\chi_k \in \widehat{\mathcal{T}\mathcal{F}}$ by

$$\chi_k(g) := \exp(2\pi\mathrm{i}\langle k, \tau(\pi(g)) \rangle) \qquad \text{for all } g \in \mathcal{T}\mathcal{F},$$

where $\pi \colon \mathcal{T}\mathcal{F} \to \mathcal{T}_{\mathcal{S}}$ is the natural surjective homomorphism.

Since $\mathcal{T}\mathcal{F}$ is a normal subgroup of $\mathcal{G}$, $\mathcal{G}$ acts on $\widehat{\mathcal{T}\mathcal{F}}$.

Lemma 2.30. *For all $g \in \mathcal{G}$ and $k, k' \in \mathbb{R}^{d_2}$ it holds*

$$\chi_k \chi_{k'} = \chi_{k+k'}$$

and

$$g \cdot \chi_k = \chi_{L(\pi(g))k},$$

where $\pi \colon \mathcal{G} \to \mathcal{S}$ is the natural surjective homomorphism.

Proof. Let $g \in \mathcal{G}$, $k, k' \in \mathbb{R}^{d_2}$ and $\pi \colon \mathcal{G} \to \mathcal{S}$ be the natural surjective homomorphism. For all $h \in \mathcal{T}\mathcal{F}$ it holds

$$\begin{aligned}
\chi_k(h)\chi_{k'}(h) &= \exp(2\pi\mathrm{i}\langle k, \tau(\pi(h)) \rangle) \exp(2\pi\mathrm{i}\langle k', \tau(\pi(h)) \rangle) \\
&= \exp(2\pi\mathrm{i}\langle k + k', \tau(\pi(h)) \rangle) \\
&= \chi_{k+k'}(h)
\end{aligned}$$

and

$$(g \cdot \chi_k)(h) = \chi_k(g^{-1}hg)$$
$$= \exp(2\pi i \langle k, \tau(\pi(g^{-1}hg)) \rangle)$$
$$= \exp(2\pi i \langle k, L(\pi(g^{-1}))\tau(\pi(h)) \rangle)$$
$$= \exp(2\pi i \langle L(\pi(g))k, \tau(\pi(h)) \rangle)$$
$$= \chi_{L(\pi(g))k}(h). \qquad \square$$

Lemma 2.31. *For all $n \in \mathbb{N}$ it holds*

$$L_S^*/n = \{k \in \mathbb{R}^{d_2} \mid \chi_k|_{\mathcal{T}^n} = 1\}.$$

Proof. Let $n \in \mathbb{N}$ and $\pi \colon \mathcal{TF} \to \mathcal{T}_S$ be the natural surjective homomorphism. First we show that $L_S^*/n \subset \{k \in \mathbb{R}^{d_2} \mid \chi_k|_{\mathcal{T}^n} = 1\}$. Let $k \in L_S^*/n$. For all $t \in \mathcal{T}$ it holds $\tau(\pi(t^n)) = n\tau(\pi(t))$ and thus,

$$\chi_k(t^n) = \exp(2\pi i \langle k, \tau(\pi(t^n)) \rangle) = \exp(2\pi i \langle nk, \tau(\pi(t)) \rangle) = 1.$$

Now we show that $\{k \in \mathbb{R}^{d_2} \mid \chi_k|_{\mathcal{T}^n} = 1\} \subset L_S^*/n$. Let $k \in \mathbb{R}^{d_2}$ such that $\chi_k|_{\mathcal{T}^n} = 1$. Let $x \in L_S$. There exists some $t \in \mathcal{T}$ such that $x = \tau(\pi(t))$. We have

$$\langle nk, x \rangle = \langle nk, \tau(\pi(t)) \rangle = \langle k, \tau(\pi(t^n)) \rangle \in \mathbb{Z},$$

where we used that $\chi_k(t^n) = 1$ in the last step. Since $x \in L_S$ was arbitrary, we have $k \in L_S^*/n$. $\qquad \square$

Definition 2.32. We define the relation $\sim$ on $\widehat{\mathcal{TF}}$ by

$$(\rho \sim \rho') :\Longleftrightarrow (\exists g \in \mathcal{G} \, \exists k \in \mathbb{R}^{d_2} : g \cdot \rho = \chi_k \rho').$$

Remark 2.33. One can also define an equivalence relation $\sim$ on the set of all representations of $\mathcal{TF}$ by

$$(\rho \sim \rho') :\Longleftrightarrow ([\rho] \sim [\rho']) \qquad \text{for all representations } \rho, \rho' \text{ on } \mathcal{TF}.$$

Lemma 2.34. *The relation $\sim$ on $\widehat{\mathcal{TF}}$ is an equivalence relation.*

Proof. It is clear that $\sim$ is reflexive.
Now we show that $\sim$ is symmetric. Let $\rho, \rho' \in \widehat{\mathcal{TF}}$ such that $\rho \sim \rho'$. There exist some $g \in \mathcal{G}$ and $k \in \mathbb{R}^{d_2}$ such that $g \cdot \rho = \chi_k \rho'$. This implies

$$g^{-1} \cdot \rho' = (g^{-1} \cdot \chi_{-k})(g^{-1} \cdot (\chi_k \rho')) = \chi_{-L(\pi(g^{-1}))k}\rho,$$

where $\pi\colon \mathcal{G} \to \mathcal{S}$ is the natural surjective homomorphism.

Now we show that $\sim$ is transitive. Let $\rho, \rho', \rho'' \in \widehat{\mathcal{TF}}$ such that $\rho \sim \rho'$ and $\rho' \sim \rho''$. There exist some $g, g' \in \mathcal{G}$ and $k, k' \in \mathbb{R}^{d_2}$ such that $g \cdot \rho = \chi_k \rho'$ and $g' \cdot \rho' = \chi_{k'} \rho''$. This implies

$$(g'g) \cdot \rho = g' \cdot (\chi_k \rho') = \chi_{L(\pi(g'))k + k'} \rho'',$$

where $\pi\colon \mathcal{G} \to \mathcal{S}$ is the natural surjective homomorphism. $\qquad\square$

Definition 2.35. For all groups $\mathcal{H} \le \mathcal{G}$ and $N \in M_0$ such that T^N is a normal subgroup of $\mathcal{H}$, let $\mathcal{H}_N$ denote the quotient group $\mathcal{H}/T^N$.

The following lemma gives an algorithm how we can determine a representation set of $\widehat{\mathcal{TF}}/\sim$.

Lemma 2.36. *Let* $m \in \mathbb{N}$ *such that* $M_0 = m\mathbb{N}$.

(i) *Every representation set of* $\{\rho \in \widehat{\mathcal{TF}} \mid \rho|_{T^m} = I_{d_\rho}\}/\sim$ *is a representation set of* $\widehat{\mathcal{TF}}/\sim$.

(ii) *The map*

$$\widehat{(\mathcal{TF})_m} \to \{\rho \in \widehat{\mathcal{TF}} \mid \rho|_{T^m} = I_{d_\rho}\}, \quad \rho \mapsto \rho \circ \pi$$

where $\pi\colon \mathcal{TF} \to (\mathcal{TF})_m$ *is the natural surjective homomorphism, is bijective. In particular, the set* $\{\rho \in \widehat{\mathcal{TF}} \mid \rho|_{T^m} = I_{d_\rho}\}$ *is finite.*

(iii) *Let* K *be a representation set of* $(L_{\mathcal{S}}^*/m)/L_{\mathcal{S}}^*$ *and* $\mathcal{P}$ *be a representation set of* $\mathcal{G}/\mathcal{TF}$. *Then, for all* $\rho, \rho' \in \{\tilde{\rho} \in \widehat{\mathcal{TF}} \mid \tilde{\rho}|_{T^m} = I_{d_{\tilde{\rho}}}\}$ *it holds*

$$(\rho \sim \rho') \iff (\exists g \in \mathcal{P}\, \exists k \in K : g \cdot \rho = \chi_k \rho').$$

Proof. Let $m \in \mathbb{N}$ such that $M_0 = m\mathbb{N}$.

(i) Let R be a representation set of $\{\rho \in \widehat{\mathcal{TF}} \mid \rho|_{T^m} = I_{d_\rho}\}/\sim$. We have to show that for all $\rho \in \widehat{\mathcal{TF}}$ there exists some $\rho' \in R$ such that $\rho \sim \rho'$. Let $\rho \in \widehat{\mathcal{TF}}$. By Proposition 2.15 the group T^m is a subgroup of the center of $\mathcal{TF}$ and thus, by Proposition B.1 for all $t \in T^m$ there exists some $\lambda \in \mathbb{C}$ such that $|\lambda| = 1$ and $\rho(t) = \lambda I_{d_\rho}$. Hence, there exists some one-dimensional representation $\chi \in \widehat{T^m}$ such that $\rho|_{T^m} = \chi I_{d_\rho}$.

There exists some $k \in \mathbb{R}^{d_2}$ such that $\chi|_{T^m} = \chi_k|_{T^m}$: By Lemma 2.12 the group T^m is isomorphic to $\mathbb{Z}^{d_2}$. Thus, there exist $t_1, \ldots, t_{d_2} \in T^m$ such that $\{t_1, \ldots, t_{d_2}\}$ generates T^m. For all $j \in \{1, \ldots, d_2\}$ there exists some

$\alpha_j \in \mathbb{R}$ such that $\exp(2\pi i \alpha_j) = \chi(t_j)$. For all $i \in \{1, \ldots, d_2\}$ let $b_i \in \mathbb{R}^{d_2}$ such that

$$\langle b_i, \tau(\pi(t_j)) \rangle = \delta_{ij} \qquad \text{for all } j \in \{1, \ldots, d_2\},$$

where $\pi \colon \mathcal{TF} \to \mathcal{T}_{\mathcal{S}}$ is the natural surjective homomorphism. For $k = \sum_{i=1}^{d_2} \alpha_i b_i \in \mathbb{R}^{d_2}$ it holds $\chi|_{\mathcal{T}^m} = \chi_k|_{\mathcal{T}^m}$.

Thus, we have $\rho|_{\mathcal{T}^m} = \chi_k|_{\mathcal{T}^m} I_{d_\rho}$. Since $\chi_{-k}\rho \in \widehat{\mathcal{TF}}$ and $(\chi_{-k}\rho)|_{\mathcal{T}^m} = I_{d_\rho}$, there exists some $\rho' \in R$ such that $\chi_{-k}\rho \sim \rho'$. There exist some $g \in \mathcal{G}$ and $l \in \mathbb{R}^{d_2}$ such that $g \cdot \rho' = \chi_l(\chi_{-k}\rho)$. This implies $\rho \sim \rho'$.

(ii) This is clear by Proposition B.2 and Remark 2.18(ii).

(iii) Let $\rho, \rho' \in \widehat{\mathcal{TF}}$ such that $\rho|_{\mathcal{T}^m} = I_{d_\rho}$, $\rho'|_{\mathcal{T}^m} = I_{d_{\rho'}}$ and $\rho \sim \rho'$. There exist some $g \in \mathcal{G}$ and $k \in \mathbb{R}^{d_2}$ such that $g \cdot \rho = \chi_k \rho'$. Let $h \in \mathcal{P}$ such that $g\mathcal{TF} = h\mathcal{TF}$. It holds $I_{d_\rho} = (g \cdot \rho)|_{\mathcal{T}^m} = (\chi_k\rho')|_{\mathcal{T}^m} = \chi_k|_{\mathcal{T}^m} I_{d_{\rho'}}$. This implies $\chi_k|_{\mathcal{T}^m} = 1$ and thus, $k \in (L_{\mathcal{S}}^*/m)$ by Lemma 2.31. Let $l \in K$ such that $lL_{\mathcal{S}}^* = kL_{\mathcal{S}}^*$. We have

$$h \cdot \rho = g \cdot \rho = \chi_k \rho' = \chi_l \rho',$$

where we used Lemma 2.22 in the first step and that $\chi_{k-l} = 1$ since $k - l \in L_{\mathcal{S}}^*$ in the last step.

The other direction of the assertion is trivial. $\qquad\qquad\square$

Corollary 2.37. *The set $\widehat{\mathcal{TF}}/\sim$ is finite.*

Proof. This is clear by Lemma 2.36. $\qquad\qquad\square$

Definition 2.38. For all $\rho \in \widehat{\mathcal{TF}}$ we define the set

$$\mathcal{G}_\rho := \left\{ (L(\pi(g)), k) \,\middle|\, g \in \mathcal{G}, k \in \mathbb{R}^{d_2} : g \cdot \rho = \chi_k \rho \right\} \subset \mathrm{E}(d_2),$$

where $\pi \colon \mathcal{G} \to \mathcal{S}$ is the natural surjective homomorphism.

Proposition 2.39. *For all $\rho \in \widehat{\mathcal{TF}}$ the set $\mathcal{G}_\rho$ is a space group and it holds*

$$L_{\mathcal{S}}^* \leq \left\{ k \in \mathbb{R}^{d_2} \,\middle|\, (I_{d_2}, k) \in \mathcal{G}_\rho \right\} \leq L_{\mathcal{S}}^*/m,$$

where $m \in \mathbb{N}$ is such that $M_0 = m\mathbb{N}$.

Proof. Let $\rho \in \widehat{\mathcal{TF}}$ and $m \in \mathbb{N}$ such that $M_0 = m\mathbb{N}$. First we show that $\mathcal{G}_\rho$ is a subgroup of $\mathrm{E}(d_2)$. Let $g_1, g_2 \in \mathcal{G}_\rho$. We have to show that $g_1 g_2^{-1} \in \mathcal{G}_\rho$. Let $\pi \colon \mathcal{G} \to \mathcal{S}$ be the natural surjective homomorphism. For all $i \in \{1, 2\}$ let $h_i \in \mathcal{G}$ and $k_i \in \mathbb{R}^{d_2}$ such that $g_i = (L(\pi(h_i)), k_i)$ and $h_i \cdot \rho = \chi_{k_i} \rho$. It holds

$$(h_1 h_2^{-1}) \cdot \rho = h_1 \cdot (h_2^{-1} \cdot \rho) = h_1 \cdot ((h_2^{-1} \cdot \chi_{-k_2}) \rho)$$
$$= ((h_1 h_2^{-1}) \cdot \chi_{-k_2})(h_1 \cdot \rho) = \chi_{k_1 - L(\pi(h_1 h_2^{-1})) k_2} \rho$$

and thus,

$$g_1 g_2^{-1} = (L(\pi(h_1 h_2^{-1})), k_1 - L(\pi(h_1 h_2^{-1})) k_2) \in \mathcal{G}_\rho.$$

Let

$$\mathcal{H} := \mathcal{G}_\rho \cap \mathrm{Trans}(d_2)$$

be the group of all translations of $\mathcal{G}_\rho$. It is clear that $\tau(\mathcal{H}) = \{k \in \mathbb{R}^{d_2} \mid (I_{d_2}, k) \in \mathcal{G}_\rho\}$.

Now we show that $\tau(\mathcal{H}) \leq L_{\mathcal{S}}^*/m$. Let $k \in \tau(\mathcal{H})$, i.e. $(I_{d_2}, k) \in \mathcal{G}_\rho$. There exists some $g \in \mathcal{G}$ such that $g \cdot \rho = \chi_k \rho$ and $L(\pi(g)) = I_{d_2}$. The latter implies $\pi(g) \in \mathcal{T}_{\mathcal{S}}$ and thus, $g \in \mathcal{TF}$. By Lemma 2.22 we have $\rho = \chi_k \rho$. Let $\tilde{\rho}$ be a representative of ρ. There exists some $T \in \mathrm{U}(d_\rho)$ such that $T^{\mathsf{H}} \tilde{\rho}(g) T = \chi_k(g) \tilde{\rho}(g)$ for all $g \in \mathcal{TF}$. Moreover, by Proposition 2.15 the set $\mathcal{T}^m$ is a subset of the center of $\mathcal{TF}$ and hence, by Proposition B.1 $\tilde{\rho}(g)$ is a scalar multiple of I_{d_ρ} for all $g \in \mathcal{T}^m$. Hence, we have $\chi_k(g) = 1$ for all $g \in \mathcal{T}^m$ and $k \in L_{\mathcal{S}}^*/m$ by Lemma 2.31.

Now we show that $L_{\mathcal{S}}^* \leq \tau(\mathcal{H})$. Let $k \in L_{\mathcal{S}}^*$. By Lemma 2.31 we have $\chi_k|_{\mathcal{T}} = 1$. Since we also have $\chi_k|_{\mathcal{F}} = 1$, we have $\chi_k = 1$. Thus we have $\mathrm{id}_{\mathcal{G}} \cdot \rho = \chi_k \rho$ and $(I_{d_2}, k) \in \mathcal{H}$, i.e. $k \in \tau(\mathcal{H})$.

Now we show that $\mathcal{G}_\rho$ is discrete. Since $\tau(\mathcal{H})$ is a subgroup of $L_{\mathcal{S}}^*/m$, the group $\mathcal{H}$ is discrete. Since $L(\mathcal{G}_\rho)$ is a subgroup of the finite group $L(\mathcal{S})$, the index $|\mathcal{G}_\rho : \mathcal{H}| = |L(\mathcal{G}_\rho)|$ is finite and thus, by [47, Theorem 7.1] the group $\mathcal{G}_\rho$ is discrete. Since $L_{\mathcal{S}}^*$ is a subgroup of $\tau(\mathcal{H})$, the group $\mathcal{G}_\rho$ contains d_2 linearly independent translations. By [18, Lemma 3, p. 415] the group $\mathcal{G}_\rho$ is a space group. $\qquad \square$

Lemma 2.40. *For all $N \in M_0$ and $\rho \in \widehat{\mathcal{TF}}$ such that $\rho|_{\mathcal{T}^N} = I_{d_\rho}$, the set $L_{\mathcal{S}}^*/N$ is invariant under $\mathcal{G}_\rho$, i.e. $\{g \cdot k \mid g \in \mathcal{G}_\rho, k \in L_{\mathcal{S}}^*/N\} = L_{\mathcal{S}}^*/N$.*

Proof. Let $N \in M_0$ and $\rho \in \widehat{\mathcal{TF}}$ such that $\rho|_{\mathcal{T}^N} = I_{d_\rho}$. Let $k \in L_{\mathcal{S}}^*/N$ and $g \in \mathcal{G}_\rho$. We have to show that $g \cdot k \in L_{\mathcal{S}}^*/N$. Let $\pi: \mathcal{G} \to \mathcal{S}$ be the natural surjective homomorphism. There exist some $h \in \mathcal{G}$ and $l \in \mathbb{R}^{d_2}$ such that $g = (L(\pi(h)), l)$ and $h \cdot \rho = \chi_l \rho$. Since $\rho|_{\mathcal{T}^N} = I_{d_\rho} = (h \cdot \rho)|_{\mathcal{T}^N}$, we have $\chi_l|_{\mathcal{T}^N} = 1$. We have

$$\chi_{g \cdot k} = \chi_{L(\pi(h)) k + l} = (h \cdot \chi_k) \chi_l$$

and thus, $\chi_{g \cdot k}|_{T^N} = 1$. By Lemma 2.31 we have $g \cdot k \in L_S^*/N$. $\qquad \square$

Definition 2.41. Let $\mathcal{H}$ be a subgroup of $\mathrm{E}(n)$. Then the set of all orbits of $\mathbb{R}^n$ under the action of $\mathcal{H}$ is written as $\mathbb{R}^n/\mathcal{H}$ and is called the *quotient of the action* or *orbit space*.

Remark 2.42. If a group $\mathcal{H} < \mathrm{E}(n)$ is discrete, then the quotient space $\mathbb{R}^n/\mathcal{H}$ equipped with the *orbit space distance function*

$$\mathbb{R}^n/\mathcal{H} \times \mathbb{R}^n/\mathcal{H} \to [0, \infty), \quad (x, y) \mapsto \mathrm{dist}(x, y)$$

is a metric space whose topology is equal to the *quotient topology*, see, e. g., [49, §6.6].

Theorem 2.43. *Let R be a representation set of $\widehat{\mathcal{TF}}/\sim$. Then, the map*

$$\bigsqcup_{\rho \in R} \mathbb{R}^{d_2}/\mathcal{G}_\rho \to \mathrm{Ind}_{\mathcal{TF}}^{\mathcal{G}}(\widehat{\mathcal{TF}})$$

$$(\mathcal{G}_\rho \cdot k, \rho) \mapsto \mathrm{Ind}_{\mathcal{TF}}^{\mathcal{G}}(\chi_k \rho),$$

where $\bigsqcup$ is the disjoint union, is bijective.

Proof. Let R be a representation set of $\widehat{\mathcal{TF}}/\sim$. We define the map

$$\varphi \colon \bigsqcup_{\rho \in R} \mathbb{R}^{d_2}/\mathcal{G}_\rho \to \mathrm{Ind}(\widehat{\mathcal{TF}})$$

$$(\mathcal{G}_\rho \cdot k, \rho) \mapsto \mathrm{Ind}(\chi_k \rho).$$

First we show that φ is well-defined. Let $\rho \in R$, $k, k' \in \mathbb{R}^{d_2}$ and $g \in \mathcal{G}_\rho$ such that $k' = g \cdot k$. Let $\pi \colon \mathcal{G} \to \mathcal{S}$ be the natural surjective homomorphism. There exist some $h \in \mathcal{G}$ and $l \in \mathbb{R}^{d_2}$ such that $g = (L(\pi(h)), l)$ and $h \cdot \rho = \chi_l \rho$. We have

$$h \cdot (\chi_k \rho) = (h \cdot \chi_k)(h \cdot \rho) = \chi_{L(\pi(h))k+l} \rho = \chi_{k'} \rho$$

and thus, $\mathrm{Ind}(\chi_k \rho) = \mathrm{Ind}(\chi_{k'} \rho)$ by Proposition 2.25.
Now we show that φ is injective. Let $\rho, \rho' \in \mathcal{R}$ and $k, k' \in \mathbb{R}^{d_2}$ such that $\mathrm{Ind}(\chi_k \rho) = \mathrm{Ind}(\chi_{k'} \rho')$. We have to show that $\rho = \rho'$ and $\mathcal{G}_\rho \cdot k = \mathcal{G}_{\rho'} \cdot k'$. By Proposition 2.25 there exists some $g \in \mathcal{G}$ such that $g \cdot (\chi_k \rho) = \chi_{k'} \rho'$. This is equivalent to $g \cdot \rho = \chi_{k'-L(\pi(g))k} \rho'$, which implies $\rho \sim \rho'$ and thus, $\rho = \rho'$. This implies that $(L(\pi(g)), k' - L(\pi(g))k) \in \mathcal{G}_\rho$ and thus,

$$\mathcal{G}_\rho \cdot k = \mathcal{G}_{\rho'} \cdot \big((L(\pi(g)), k' - L(\pi(g))k) \cdot k \big) = \mathcal{G}_{\rho'} \cdot k'.$$

Now we show that φ is surjective. Let $\rho \in \widehat{T\mathcal{F}}$. Let $\rho' \in R$ such that $\rho \sim \rho'$. There exist some $g \in \mathcal{G}$ and $k \in \mathbb{R}^{d_2}$ such that $g \cdot \rho = \chi_k \rho'$. By Proposition 2.25 we have

$$\varphi((\mathcal{G}_{\rho'} \cdot k, \rho')) = \mathrm{Ind}(\chi_k \rho') = \mathrm{Ind}(g \cdot \rho) = \mathrm{Ind}\,\rho. \qquad \square$$

Corollary 2.44. *Let R be a representation set of $\{\rho \in \widehat{T\mathcal{F}} \mid \rho|_{T^m} = I_{d_\rho}\}/\sim$, where $m \in \mathbb{N}$ is such that $M_0 = m\mathbb{N}$. Then the maps*

(i) $\displaystyle\bigsqcup_{\rho \in R} \{k/N \mid k \in L_\mathcal{S}^*, N \in M_0\}/\mathcal{G}_\rho$

$$\to \mathrm{Ind}(\{\rho \in \widehat{T\mathcal{F}} \mid \exists N \in M_0 : \rho|_{T^N} = I_{d_\rho}\})$$
$$(\mathcal{G}_\rho \cdot (k/N), \rho) \mapsto \mathrm{Ind}(\chi_{k/N}\rho)$$

(ii) $\displaystyle\bigsqcup_{\rho \in R} (L_\mathcal{S}^*/N)/\mathcal{G}_\rho \to \mathrm{Ind}(\{\rho \in \widehat{T\mathcal{F}} \mid \rho|_{T^N} = I_{d_\rho}\})$

$$(\mathcal{G}_\rho \cdot k, \rho) \mapsto \mathrm{Ind}(\chi_k \rho),$$

where $\bigsqcup$ is the disjoint union, $\mathrm{Ind} = \mathrm{Ind}_{T\mathcal{F}}^{\mathcal{G}}$ and $N \in M_0$ in (ii) is arbitrary, are bijective.

Proof. Let $m \in \mathbb{N}$ such that $M_0 = m\mathbb{N}$ and R be a representation set of $\{\rho \in \widehat{T\mathcal{F}} \mid \rho|_{T^m} = I_{d_\rho}\}/\sim$. By Lemma 2.36 the set R is a representation set of $\widehat{T\mathcal{F}}/\sim$.
(i) We define the map

$$\psi \colon \bigsqcup_{\rho \in R} \{k/N \mid k \in L_\mathcal{S}^*, N \in M_0\}/\mathcal{G}_\rho$$
$$\to \mathrm{Ind}(\{\rho \in \widehat{T\mathcal{F}} \mid \exists N \in M_0 : \rho|_{T^N} = I_{d_\rho}\})$$
$$(\mathcal{G}_\rho \cdot (k/N), \rho) \mapsto \mathrm{Ind}(\chi_{k/N}\rho).$$

First we show that ψ is well-defined. Let $\rho \in R$, $k \in L_\mathcal{S}^*$ and $N \in M_0$. Since $T^N \subset T^m$ and by Lemma 2.31, we have $(\chi_{k/N}\rho)|_{T^N} = I_{d_\rho}$. By Lemma 2.40 for all $N \in M_0$ we have $(L_\mathcal{S}^*/N)/\mathcal{G}_\rho \subset \mathbb{R}^{d_2}/\mathcal{G}_\rho$ and thus, by Theorem 2.43 the map ψ is well-defined.
Since the map of Theorem 2.43 is injective, also ψ is injective.
It remains to show that ψ is surjective. Let $\rho \in \widehat{T\mathcal{F}}$ and $N \in M_0$ such that $\rho|_{T^N} = I_{d_\rho}$. There exists some $\rho' \in R$ such that $\rho \sim \rho'$. There exist some $g \in \mathcal{G}$ and $k \in \mathbb{R}^{d_2}$ such that $g \cdot \rho = \chi_k \rho'$. We have

$(g \cdot \rho)|_{\mathcal{T}^N} = I_{d_\rho} = \rho'|_{\mathcal{T}^N}$ and thus, $\chi_k|_{\mathcal{T}^N} = 1$. By Lemma 2.31 we have $k \in L_{\mathcal{S}}^* / N$ and thus

$$\psi((\mathcal{G}_{\rho'} \cdot k, \rho')) = \mathrm{Ind}(\chi_k \rho') = \mathrm{Ind}(g \cdot \rho) = \mathrm{Ind}\,\rho,$$

by Proposition 2.25. (ii) The proof is analogous to the proof of (i). □

2.3.3. Harmonic analysis

Definition 2.45. Let S be a set and $N \in M_0$. A function $u \colon \mathcal{G} \to S$ is called $\mathcal{T}^N$-*periodic* if

$$u(g) = u(gt) \qquad \text{for all } g \in \mathcal{G} \text{ and } t \in \mathcal{T}^N.$$

A function $u \colon \mathcal{G} \to S$ is called *periodic* if there exists some $N \in M_0$ such that u is $\mathcal{T}^N$-periodic.

We equip $\mathbb{C}^{m \times n}$ with the inner product $\langle \cdot, \cdot \rangle$ defined by

$$\langle A, B \rangle := \sum_{i=1}^{m} \sum_{j=1}^{n} a_{ij} \overline{b_{ij}} \qquad \text{for all } A, B \in \mathbb{C}^{m \times n}$$

and let $\| \cdot \|$ denote the induced norm. We define the set

$$L_{\mathrm{per}}^{\infty}(\mathcal{G}, \mathbb{C}^{m \times n}) := \{ u \colon \mathcal{G} \to \mathbb{C}^{m \times n} \mid u \text{ is periodic} \}.$$

Remark 2.46. (i) The inner product $\langle \cdot, \cdot \rangle$ on $\mathbb{C}^{m \times n}$ is the Frobenius inner product.

(ii) If $\mathcal{G}$ is finite and S a set, then every function from $\mathcal{G}$ to S is periodic and in particular, we have $L_{\mathrm{per}}^{\infty}(\mathcal{G}, \mathbb{C}^{m \times n}) = \{ u \colon \mathcal{G} \to \mathbb{C}^{m \times n} \}$.

The following Lemma shows that the above definition of periodicity is independent of the choice of $\mathcal{T}$.

Lemma 2.47. *Let S be a set. A function $u \colon \mathcal{G} \to S$ is periodic if and only if there exists some $N \in \mathbb{N}$ such that*

$$u(g) = u(gh) \qquad \text{for all } g \in \mathcal{G} \text{ and } h \in \mathcal{G}^N.$$

Proof. Let S be a set and $u \colon \mathcal{G} \to S$ be $\mathcal{T}^N$-periodic for some $N \in M_0$. By Theorem 2.17 the function u is $\mathcal{T}^{|\mathcal{F}|N}$-periodic. By Proposition 2.15 it holds

$$\mathcal{G}^{|\mathcal{G}/\mathcal{T}\mathcal{F}||\mathcal{F}|N} \subset (\mathcal{T}\mathcal{F})^{|\mathcal{F}|N} \subset (\mathcal{T}^N \mathcal{F})^{|\mathcal{F}|} = \mathcal{T}^{|\mathcal{F}|N} \mathcal{F}^{|\mathcal{F}|} = \mathcal{T}^{|\mathcal{F}|N} \subset \mathcal{T}^N.$$

and thus, we have

$$u(g) = u(gh) \qquad \text{for all } g \in \mathcal{G} \text{ and } h \in \mathcal{G}^{|\mathcal{G}/\mathcal{TF}||\mathcal{F}|N}.$$

The other direction is trivial since by Theorem 2.17 for all $N \in \mathbb{N}$ there exists some $n \in \mathbb{N}$ such that $nN \in M_0$. $\qquad\square$

The following lemma characterizes the periodic functions on $\mathcal{G}$ with the aid of the quotient groups $\mathcal{G}/\mathcal{T}^N$.

Lemma 2.48. *If $N \in M_0$ and $u \colon \mathcal{G} \to S$ is $\mathcal{T}^N$-periodic, then the function*

$$\mathcal{G}_N \to S$$
$$g\mathcal{T}^N \mapsto u(g)$$

is well-defined. Moreover, we have

$$L^\infty_{\mathrm{per}}(\mathcal{G}, \mathbb{C}^{m \times n}) = \left\{ \mathcal{G} \to \mathbb{C}^{m \times n}, g \mapsto u(g\mathcal{T}^N) \,\Big|\, N \in M_0, u \colon \mathcal{G}_N \to \mathbb{C}^{m \times n} \right\}.$$

Proof. This follows immediately from the definition of $L^\infty_{\mathrm{per}}(\mathcal{G}, \mathbb{C}^{m \times n})$. $\qquad\square$

Lemma 2.49. *The set $L^\infty_{\mathrm{per}}(\mathcal{G}, \mathbb{C}^{m \times n})$ is a vector space.*

Proof. If $u_1 \in L^\infty_{\mathrm{per}}(\mathcal{G}, \mathbb{C}^{m \times n})$ is $\mathcal{T}^{N_1}$-periodic and $u_2 \in L^\infty_{\mathrm{per}}(\mathcal{G}, \mathbb{C}^{m \times n})$ is $\mathcal{T}^{N_2}$-periodic for some $N_1, N_2 \in M_0$, then $u_1 + u_2$ is $\mathcal{T}^{N_1 N_2}$-periodic. Thus, $L^\infty_{\mathrm{per}}(\mathcal{G}, \mathbb{C}^{m \times n})$ is closed under addition. The other conditions are trivial. $\qquad\square$

Definition 2.50. For all $N \in M_0$ let $\mathcal{C}_N$ be a representation set of $\mathcal{G}/\mathcal{T}^N$.

Remark 2.51.   (i) If $\mathcal{G}$ is finite, we have $\mathcal{C}_N = \mathcal{G}$ for all $N \in M_0$.

(ii) Let $\mathcal{G}$ be infinite. There exists some $m \in \mathbb{N}$ such that $M_0 = m\mathbb{N}$ and there exist $t_1, \ldots, t_{d_2} \in \mathcal{T}^m$ such that $\{t_1, \ldots, t_{d_2}\}$ generates $\mathcal{T}^m$. Let $\mathcal{C}$ be a representation set of $\mathcal{G}/\mathcal{T}^m$. Then for all $N \in M_0$ a feasible choice for $\mathcal{C}_N$ is

$$\mathcal{C}_N = \left\{ t_1^{n_1} \ldots t_{d_2}^{n_{d_2}} g \,\Big|\, n_1, \ldots, n_{d_2} \in \{0, \ldots, N/m - 1\}, g \in \mathcal{C} \right\}.$$

For this choice, for all $x \in \mathbb{R}^d$ and large $N \in M_0$ the set $\mathcal{C}_N \cdot x$ is similar to a cube which explains the nomenclature.

We equip the vector space $L_{\mathrm{per}}^\infty(\mathcal{G}, \mathbb{C}^{m \times n})$ with an inner product.

Definition 2.52. We define the inner product $\langle \, \cdot \, , \, \cdot \, \rangle$ on $L_{\mathrm{per}}^\infty(\mathcal{G}, \mathbb{C}^{m \times n})$ by

$$\langle u, v \rangle := \frac{1}{|\mathcal{C}_N|} \sum_{g \in \mathcal{C}_N} \langle u(g), v(g) \rangle \qquad \text{if } u \text{ and } v \text{ are } \mathcal{T}^N\text{-periodic}$$

for all $u, v \in L_{\mathrm{per}}^\infty(\mathcal{G}, \mathbb{C}^{m \times n})$. We denote the induced norm by $\| \cdot \|_2$.

Definition 2.53. Let $\mathcal{E}$ be a representation set of $\{\rho \in \widehat{\mathcal{G}} \mid \rho \text{ is periodic}\}$.

Remark 2.54. (i) All representations of $\mathcal{E}$ are unitary by Definition 2.19 which is necessary for the Plancherel formula in Proposition 2.56.

(ii) For all $N \in M_0$ a representation of $\mathcal{G}$ is $\mathcal{T}^N$-periodic if and only if $\rho|_{\mathcal{T}^N} = I_{d_\rho}$.

(iii) Proposition B.2 shows that

$$\{\rho \in \widehat{\mathcal{G}} \mid \rho \text{ is periodic}\} = \{\rho \circ \pi_N \mid N \in M_0, \rho \in \widehat{\mathcal{G}_N}\},$$

where π_N is the natural surjective homomorphism from $\mathcal{G}$ to $\mathcal{G}_N$ for all $N \in M_0$.

Definition 2.55. For all $u \in L_{\mathrm{per}}^\infty(\mathcal{G}, \mathbb{C}^{m \times n})$ and for all periodic representations ρ of $\mathcal{G}$ we define

$$\widehat{u}(\rho) := \frac{1}{|\mathcal{C}_N|} \sum_{g \in \mathcal{C}_N} u(g) \otimes \rho(g) \in \mathbb{C}^{(md_\rho) \times (nd_\rho)},$$

where $N \in M_0$ is such that u and ρ are $\mathcal{T}^N$-periodic and $\otimes$ denotes the Kronecker product, see Definition D.1.

Proposition 2.56 (The Plancherel formula)**.** *The Fourier transformation*

$$\widehat{} : L_{\mathrm{per}}^\infty(\mathcal{G}, \mathbb{C}^{m \times n}) \to \bigoplus_{\rho \in \mathcal{E}} \mathbb{C}^{(md_\rho) \times (nd_\rho)}, \qquad u \mapsto (\widehat{u}(\rho))_{\rho \in \mathcal{E}}$$

is well-defined and bijective. Moreover, we have the Plancherel formula

$$\langle u, v \rangle = \sum_{\rho \in \mathcal{E}} d_\rho \langle \widehat{u}(\rho), \widehat{v}(\rho) \rangle \qquad \text{for all } u, v \in L_{\mathrm{per}}^\infty(\mathcal{G}, \mathbb{C}^{m \times n}).$$

Proof. We show that the well-known Plancherel formula for finite groups, see, e. g., [54, Theorem III.8.1], implies the Plancherel formula of the proposition. Let $N \in M_0$ and $\pi_N \colon \mathcal{G} \to \mathcal{G}_N$ be the natural surjective homomorphism. The map

$$f_1 \colon \{u \colon \mathcal{G}_N \to \mathbb{C}^{m \times n}\} \to \{u \in L_{\mathrm{per}}^\infty(\mathcal{G}, \mathbb{C}^{m \times n}) \,|\, u \text{ is } \mathcal{T}^N\text{-periodic}\}$$

$$u \mapsto u \circ \pi_N$$

is bijective. Let $\mathcal{E}_N = \{\rho \,|\, \rho \text{ is a representation of } \mathcal{G}_N, \rho \circ \pi_N \in \mathcal{E}\}$. We have $\{\rho \circ \pi_N \,|\, \rho \in \mathcal{E}_N\} = \{\rho \in \mathcal{E} \,|\, \rho \text{ is } \mathcal{T}^N\text{-periodic}\}$. Thus the map

$$f_2 \colon \bigoplus_{\rho \in \mathcal{E}, \ \rho \text{ is } \mathcal{T}^N\text{-periodic}} \mathbb{C}^{(md_\rho) \times (nd_\rho)} \to \bigoplus_{\rho \in \mathcal{E}_N} \mathbb{C}^{(md_\rho) \times (nd_\rho)}$$

$$(A_\rho)_{\rho \in \mathcal{E}, \ \rho \text{ is } \mathcal{T}^N\text{-periodic}} \mapsto (A_{\rho \circ \pi_N})_{\rho \in \mathcal{E}_N}$$

is bijective. By Proposition B.2 the set $\mathcal{E}_N$ is a representation set of $\widehat{\mathcal{G}_N}$. For all $u \colon \mathcal{G}_N \to \mathbb{C}^{m \times n}$ and $\rho \in \mathcal{E}_N$ we define $\widehat{u}(\rho) = \frac{1}{|\mathcal{G}_N|} \sum_{g \in \mathcal{G}_N} u(g) \otimes \rho(g)$. By the Plancherel formula for finite groups, see, e. g., [9, Proposition 16.16], the Fourier transformation

$$\widehat{} \colon \{u \colon \mathcal{G}_N \to \mathbb{C}^{m \times n}\} \to \bigoplus_{\rho \in \mathcal{E}_N} \mathbb{C}^{(md_\rho) \times (nd_\rho)}, \quad u \mapsto (\widehat{u}(\rho))_{\rho \in \mathcal{E}_N}$$

is bijective and it holds $\frac{1}{|\mathcal{G}_N|} \sum_{g \in \mathcal{G}_N} \langle u(g), v(g) \rangle = \sum_{\rho \in \mathcal{E}_N} d_\rho \langle \widehat{u}(\rho), \widehat{v}(\rho) \rangle$ for all $u, v \colon \mathcal{G}_N \to \mathbb{C}^{m \times n}$. The diagram

$$
\begin{array}{ccc}
\{u \in L_{\mathrm{per}}^\infty(\mathcal{G}, \mathbb{C}^{m \times n}) \,|\, u \text{ is } \mathcal{T}^N\text{-periodic}\} & \xrightarrow{\ \widehat{}\ } & \bigoplus_{\substack{\rho \in \mathcal{E} \\ \rho \text{ is } \mathcal{T}^N\text{-periodic}}} \mathbb{C}^{(md_\rho) \times (nd_\rho)} \\[2em]
\Big\uparrow {\scriptstyle f_1} & & \Big\downarrow {\scriptstyle f_2} \\[2em]
\{u \colon \mathcal{G}_N \to \mathbb{C}^{m \times n}\} & \xrightarrow{\ \widehat{}\ } & \bigoplus_{\rho \in \mathcal{E}_N} \mathbb{C}^{(md_\rho) \times (nd_\rho)}
\end{array}
$$

commutes, where the top map is defined by $u \mapsto (\widehat{u}(\rho))_{\rho \in \mathcal{E}, \rho \text{ is } \mathcal{T}^N\text{-periodic}}$. Thus, the map

$$\widehat{} \colon \{u \in L_{\mathrm{per}}^\infty(\mathcal{G}, \mathbb{C}^{m \times n}) \,|\, u \text{ is } \mathcal{T}^N\text{-periodic}\} \to \bigoplus_{\substack{\rho \in \mathcal{E} \\ \rho \text{ is } \mathcal{T}^N\text{-periodic}}} \mathbb{C}^{(md_\rho) \times (nd_\rho)}$$

$$(2.6)$$

is bijective and we have

$$\langle u, v \rangle = \sum_{\rho \in \mathcal{E},\ \rho \text{ is } \mathcal{T}^N\text{-periodic}} d_\rho \langle \widehat{u}(\rho), \widehat{v}(\rho) \rangle$$

for all $\mathcal{T}^N$-periodic functions $u, v \in L^\infty_{\mathrm{per}}(\mathcal{G}, \mathbb{C}^{m \times n})$.

Since $N \in M_0$ was arbitrary, for all $u \in L^\infty_{\mathrm{per}}(\mathcal{G}, \mathbb{C}^{m \times n})$, for all $N \in M_0$ such that u is $\mathcal{T}^N$-periodic and $n \in \mathbb{N}$ it holds

$$\sum_{\rho \in \mathcal{E},\ \rho \text{ is } \mathcal{T}^N\text{-periodic}} d_\rho \|\widehat{u}(\rho)\|^2 = \|u\|_2^2 = \sum_{\rho \in \mathcal{E},\ \rho \text{ is } \mathcal{T}^{nN}\text{-periodic}} d_\rho \|\widehat{u}(\rho)\|^2.$$
(2.7)

By (2.7) for all $u \in L^\infty_{\mathrm{per}}(\mathcal{G}, \mathbb{C}^{m \times n})$ and $N \in M_0$ such that u is $\mathcal{T}^N$-periodic, we have

$$\{\rho \in \mathcal{E} \,|\, \widehat{u}(\rho) \neq 0\} \subset \{\rho \in \mathcal{E} \,|\, \rho \text{ is } \mathcal{T}^N\text{-periodic}\}. \qquad (2.8)$$

By (2.7) and (2.8) the Fourier transformation

$$\widehat{\ }\colon L^\infty_{\mathrm{per}}(\mathcal{G}, \mathbb{C}^{m \times n}) \to \bigoplus_{\rho \in \mathcal{E}} \mathbb{C}^{(md_\rho) \times (nd_\rho)}$$

is well-defined and we have

$$\langle u, v \rangle = \sum_{\rho \in \mathcal{E}} d_\rho \langle \widehat{u}(\rho), \widehat{v}(\rho) \rangle$$

for all $u, v \in L^\infty_{\mathrm{per}}(\mathcal{G}, \mathbb{C}^{m \times n})$. Moreover, since the map defined in (2.6) is injective and $L^\infty_{\mathrm{per}}(\mathcal{G}, \mathbb{C}^{m \times n}) = \bigcup_{N \in M_0} \{u \in L^\infty_{\mathrm{per}}(\mathcal{G}, \mathbb{C}^{m \times n}) \,|\, u \text{ is } \mathcal{T}^N\text{-periodic}\}$, the Fourier transformation is injective. Analogously, the Fourier transformation is surjective. $\qquad \square$

Remark 2.57. (i) The above proof also shows that for all $u\colon \mathcal{G} \to \mathbb{C}^{m \times n}$ and $N \in M_0$ such that u is $\mathcal{T}^N$-periodic, we have

$$\{\rho \in \mathcal{E} \,|\, \widehat{u}(\rho) \neq 0\} \subset \{\rho \in \mathcal{E} \,|\, \rho \text{ is } \mathcal{T}^N\text{-periodic}\}.$$

Moreover, for all $N \in M_0$ the map

$$\{u\colon \mathcal{G} \to \mathbb{C}^{m \times n} \,|\, u \text{ is } \mathcal{T}^N\text{-periodic}\} \to \bigoplus_{\substack{\rho \in \mathcal{E} \\ \rho \text{ is } \mathcal{T}^N\text{-periodic}}} \mathbb{C}^{(md_\rho) \times (nd_\rho)}$$

$$u \mapsto \left(\widehat{u}(\rho) \right)$$

is bijective.

(ii) It is easy to see that by the above proposition we have also a description of the completion of $L_{\mathrm{per}}^\infty(\mathcal{G}, \mathbb{C}^{m\times n})$ with respect to the norm $\|\cdot\|_2$. We have

$$\overline{L_{\mathrm{per}}^\infty(\mathcal{G}, \mathbb{C}^{m\times n})}^{\|\cdot\|_2} = \left\{ u\colon \mathcal{G} \to \mathbb{C}^{m\times n} \;\middle|\; \sum_{\rho\in\mathcal{E}} d_\rho \|\widehat{u}(\rho)\|^2 < \infty \right\}$$

and the map

$$\overline{L_{\mathrm{per}}^\infty(\mathcal{G}, \mathbb{C}^{m\times n})}^{\|\cdot\|_2} \to \left\{ a \in \prod_{\rho\in\mathcal{E}} \mathbb{C}^{(md_\rho)\times(nd_\rho)} \;\middle|\; \sum_{\rho\in\mathcal{E}} d_\rho \|a(\rho)\|^2 < \infty \right\}$$

$$u \mapsto (\widehat{u}(\rho))_{\rho\in\mathcal{E}}$$

is bijective.

Lemma 2.58. *Let $f \in L_{\mathrm{per}}^\infty(\mathcal{G}, \mathbb{C}^{m\times n})$, $g \in \mathcal{G}$ and $\tau_g f$ denote the translated function $f(\cdot\, g)$. Then we have $\tau_g f \in L_{\mathrm{per}}^\infty(\mathcal{G}, \mathbb{C}^{m\times n})$ and*

$$\widehat{\tau_g f}(\rho) = \widehat{f}(\rho)(I_n \otimes \rho(g^{-1}))$$

for all periodic representations ρ of $\mathcal{G}$.

Proof. Let $f \in L_{\mathrm{per}}^\infty(\mathcal{G}, \mathbb{C}^{m\times n})$, $g \in \mathcal{G}$ and ρ be a periodic representation. Let $N \in M_0$ such that f and ρ are $\mathcal{T}^N$-periodic. The function $\tau_g f$ is $\mathcal{T}^N$-periodic and we have

$$\begin{aligned}
\widehat{\tau_g f}(\rho) &= \frac{1}{|\mathcal{C}_N|} \sum_{h\in\mathcal{C}_N} \tau_g f(h) \otimes \rho(h) \\
&= \frac{1}{|\mathcal{C}_N|} \sum_{h\in\mathcal{C}_N} f(hg) \otimes \rho(h) \\
&= \frac{1}{|\mathcal{C}_N|} \sum_{h\in\mathcal{C}_N} f(h) \otimes \rho(hg^{-1}) \\
&= \frac{1}{|\mathcal{C}_N|} \sum_{h\in\mathcal{C}_N} f(h) \otimes (\rho(h)\rho(g^{-1})) \\
&= \left(\frac{1}{|\mathcal{C}_N|} \sum_{h\in\mathcal{C}_N} f(h) \otimes \rho(h) \right)(I_n \otimes \rho(g^{-1})) \\
&= \widehat{f}(\rho)(I_n \otimes \rho(g^{-1})),
\end{aligned}$$

where in the third step we made a substitution and used that $\mathcal{C}_N$ and $\mathcal{C}_N g$ are representation sets of $\mathcal{G}/\mathcal{T}^N$ and that the function $h \mapsto f(h) \otimes \rho(hg^{-1})$ is $\mathcal{T}^N$-periodic. $\qquad\square$

Definition 2.59. For all $u \in L^1(\mathcal{G}, \mathbb{C}^{m \times n})$ and all representations ρ of $\mathcal{G}$ we define

$$\widehat{u}(\rho) := \sum_{g \in \mathcal{G}} u(g) \otimes \rho(g).$$

Remark 2.60. If the group $\mathcal{G}$ is finite, ρ is a representation of $\mathcal{G}$ and $u \in L^1(\mathcal{G}, \mathbb{C}^{m \times n}) = L^\infty_{\mathrm{per}}(\mathcal{G}, \mathbb{C}^{m \times n})$, then the Definitions 2.55 and 2.59 for $\widehat{u}(\rho)$ differ by the multiplicative constant $|\mathcal{G}|$, but it will always be clear from the context which of the both definitions is meant. If $\mathcal{G}$ is infinite, then $L^1(\mathcal{G}, \mathbb{C}^{m \times n}) \cap L^\infty_{\mathrm{per}}(\mathcal{G}, \mathbb{C}^{m \times n}) = \{0\}$ and thus, there is no ambiguity.

Definition 2.61. For all $u \in L^1(\mathcal{G}, \mathbb{C}^{l \times m})$ and $v \in L^\infty_{\mathrm{per}}(\mathcal{G}, \mathbb{C}^{m \times n})$ we define the convolution $u * v \in L^\infty_{\mathrm{per}}(\mathcal{G}, \mathbb{C}^{l \times n})$ by

$$u * v(g) := \sum_{h \in \mathcal{G}} u(h) v(h^{-1} g) \qquad \text{for all } g \in \mathcal{G}.$$

Lemma 2.62. *Let $u \in L^1(\mathcal{G}, \mathbb{C}^{l \times m})$, $v \in L^\infty_{\mathrm{per}}(\mathcal{G}, \mathbb{C}^{m \times n})$ and ρ be a periodic representation of $\mathcal{G}$. Then*

(i) *the convolution $u * v$ is $\mathcal{T}^N$-periodic if v is $\mathcal{T}^N$-periodic and*

(ii) *we have*

$$\widehat{u * v}(\rho) = \widehat{u}(\rho)\widehat{v}(\rho).$$

Proof. Let $u \in L^1(\mathcal{G}, \mathbb{C}^{l \times m})$, $v \in L^\infty_{\mathrm{per}}(\mathcal{G}, \mathbb{C}^{m \times n})$ and ρ be a periodic representation of $\mathcal{G}$. Let $N \in M_0$ such that v and ρ are $\mathcal{T}^N$-periodic. By Definition 2.61 it is clear that $u * v$ is $\mathcal{T}^N$-periodic and thus we have $u * v \in L^\infty_{\mathrm{per}}(\mathcal{G}, \mathbb{C}^{m \times n})$ as claimed in Definition 2.61. We have

$$\widehat{u * v}(\rho) = \frac{1}{|\mathcal{C}_N|} \sum_{g \in \mathcal{C}_N} u * v(g) \otimes \rho(g)$$

$$= \frac{1}{|\mathcal{C}_N|} \sum_{g \in \mathcal{C}_N} \sum_{h \in \mathcal{G}} \big(u(h) v(h^{-1} g)\big) \otimes \rho(g)$$

$$= \frac{1}{|\mathcal{C}_N|} \sum_{g \in \mathcal{C}_N} \sum_{h \in \mathcal{G}} \big(u(h) \otimes \rho(h)\big)\big(v(h^{-1} g) \otimes \rho(h^{-1} g)\big)$$

$$= \left(\sum_{h \in \mathcal{G}} u(h) \otimes \rho(h) \right) \left(\frac{1}{|\mathcal{C}_N|} \sum_{g \in \mathcal{C}_N} v(g) \otimes \rho(g) \right)$$

$$= \widehat{u}(\rho) \widehat{v}(\rho). \qquad \qquad \square$$

2.3.4. The Cauchy-Born rule

The Cauchy-Born rule generalizes in a natural way to objective structures, see [41]. The generalization postulates that if an objective structure is subjected to a (small) linear macroscopic deformation, all atoms
will follow the deformation still forming an objective structure. Thus,
if the Cauchy-Born rule holds, for each linear macroscopic deformation,
there exists an appropriate group which describes the objective structure.

Definition 2.63. Suppose that $L(\mathcal{S}) = \{I_{d_2}\}$ or $L(\mathcal{S}) = \{I_{d_2}, -I_{d_2}\}$.
Then, for all transformation matrices $A \in \mathrm{GL}(d_2)$ we define the group

$$\mathcal{G}_A = \left\{ \left(B, \begin{pmatrix} I_{d_1} & 0 \\ 0 & A \end{pmatrix} b \right) \middle| (B, b) \in \mathcal{G} \right\}.$$

It is easy to see that the group $\mathcal{G}$ is isomorphic to $\mathcal{G}_A$ and the natural
isomorphism is given by $(B, b) \mapsto (B, (I_{d_1} \oplus A)b)$. Moreover, the group
$\mathcal{G}_A$ is also a discrete subgroup of $\mathrm{E}(d)$.

Remark 2.64.   (i) The center of $\mathrm{O}(d_2)$ is $\{I_{d_2}, -I_{d_2}\}$.

(ii) Notice that the premise $L(\mathcal{S}) = \{I_{d_2}\}$ or $L(\mathcal{S}) = \{I_{d_2}, -I_{d_2}\}$ is
necessary since for an arbitrary $\mathcal{G}$ and $A \in \mathrm{GL}(d_2)$ the set

$$\left\{ \left(B, \begin{pmatrix} I_{d_1} & 0 \\ 0 & A \end{pmatrix} b \right) \middle| (B, b) \in \mathcal{G} \right\}$$

is not a group in general. Also if we assume $\mathrm{rank}(A - I_{d_2}) = 1$, the
set $\mathcal{G}_A$ is not a group in general, see Example 2.65.

Example 2.65. In this example we present a set $S \subset \mathbb{R}^2$ and two discrete
groups $\mathcal{G}_1, \mathcal{G}_2 < \mathrm{E}(2)$ such that S is the orbit of the two groups, and such
that the group $(\mathcal{G}_1)_A$ is well-defined, but the term $(\mathcal{G}_2)_A$ is in general not
meaningful for $A \in \mathrm{GL}(d_2)$.
Let $x = (1/2, 1/2) \in \mathbb{R}^2$ and $S = x + \mathbb{Z}^2$. Let $t_1 = (I_2, e_1)$, $t_2 = (I_2, e_2)$
and $\mathcal{G}_1 = \langle t_1, t_2 \rangle$. Let $s_1 = (I_2, 2e_1)$, $s_2 = (I_2, 2e_2)$, $p = (R(\pi/2), 0)$ and
$\mathcal{G}_2 = \langle s_1, s_2, p \rangle$. Then $S = \mathcal{G}_1 \cdot x$ and $S = \mathcal{G}_2 \cdot x$. For all $A \in \mathrm{GL}(2)$ the
group $(\mathcal{G}_1)_A$ is well-defined, but for e. g. $A = \left(\begin{smallmatrix} 1 & 0 \\ 0 & 2 \end{smallmatrix} \right)$ it is not possible to
define a group $(\mathcal{G}_2)_A$ in this way.

2.3.5. A representation of quotient groups as semidirect products

By Definition 2.13 for all $m \in M_0$ the group $\mathcal{T}^m$ is a normal subgroup of $\mathcal{G}$, but in general there does not exist any group $\mathcal{H} < \mathcal{G}$ such that $\mathcal{G} = \mathcal{T}^m \rtimes \mathcal{H}$, see Example 2.66. In this section we determine for $m \in M_0$ and appropriate $N \in m\mathbb{N}$ a group $\mathcal{H} \leq \mathcal{G}/\mathcal{T}^N$ such that

$$\mathcal{G}/\mathcal{T}^N = \mathcal{T}^m/\mathcal{T}^N \rtimes \mathcal{H},$$

see Theorem 2.72. The proof is similar to the proof of the Schur-Zassenhaus theorem, see, e. g., [3]. If $\mathcal{G}$ is a space group, for appropriate $N \in \mathbb{N}$ the existence of a group $\mathcal{H}$ such that $\mathcal{G}/\mathcal{T}^N = \mathcal{T}/\mathcal{T}^N \rtimes \mathcal{H}$ is mentioned in [7, p. 299] and in [29, p. 376].

Example 2.66 (Symmorphic and nonsymmorphic space groups). Here we give the definition of a symmorphic and a nonsymmorphic space group. For both of these groups we give an example.
Let $\mathcal{G}$ be a space group and $\mathcal{T}$ its subgroup of translations. If there exists a group $\mathcal{H} < \mathcal{G}$ such that $\mathcal{G} = \mathcal{T} \rtimes \mathcal{H}$, then $\mathcal{G}$ is said to be a *symmorphic* space group, see e. g., [47, Section 9.1]. Otherwise, $\mathcal{G}$ is a *nonsymmorphic* space group.
Let $d = 2$, $t_1 = (I_2, e_1)$, $t_2 = (I_2, e_2)$, $id = (I_2, 0)$, $p_1 = \left(\left(\begin{smallmatrix} 1 & 0 \\ 0 & -1 \end{smallmatrix} \right), 0 \right)$ and $p_2 = \left(\left(\begin{smallmatrix} 1 & 0 \\ 0 & -1 \end{smallmatrix} \right), \left(\begin{smallmatrix} 0,5 \\ 0 \end{smallmatrix} \right) \right)$. The space group

$$\left\{ tp \, \middle| \, t \in \langle t_1, t_2 \rangle, p \in \{id, p_1\} \right\} < \mathrm{E}(2)$$

is symmorphic and equal to $\mathcal{T} \rtimes \mathcal{H}$ with $\mathcal{T} = \langle t_1, t_2 \rangle$ and $\mathcal{H} = \langle p_1 \rangle$. The space group

$$\left\{ tp \, \middle| \, t \in \langle t_1, t_2 \rangle, p \in \{id, p_2\} \right\} < \mathrm{E}(2)$$

is nonsymmorphic, since it does not contain any element of order 2, but the order of the quotient group of the space group by its subgroup of all translations is 2.

Definition 2.67. Let $\tilde{\tau} \colon L(\mathcal{S}) \to \tau(\mathcal{S})$ be a map such that $(P, \tilde{\tau}(P)) \in \mathcal{S}$ for all $P \in L(\mathcal{S})$. We define the map

$$\bar{\tau} \colon L(\mathcal{S}) \times L(\mathcal{S}) \to \tau(\mathcal{T}_{\mathcal{S}})$$
$$(P, Q) \mapsto \tilde{\tau}(P) + P\tilde{\tau}(Q) - \tilde{\tau}(PQ).$$

Furthermore, for all $n \in \mathbb{N}$ coprime to $|L(\mathcal{S})|$ we define the set

$$\mathcal{P}_{\mathcal{S}}^{(n)} := \left\{ \left(P, \tilde{\tau}(P) - a(n) \sum_{Q \in L(\mathcal{S})} \bar{\tau}(P, Q) \right) \, \middle| \, P \in L(\mathcal{S}) \right\} \subset \mathcal{S},$$

where $a(n) = \max\{\tilde{a} \in \{0, -1, \dots\} \mid \exists b \in \mathbb{Z} \text{ such that } \tilde{a}|L(\mathcal{S})| + bn = 1\}$.
For all $n \in \mathbb{N}$ coprime to $|L(\mathcal{S})|$ let $\mathcal{P}^{(n)} \subset \mathcal{G}$ be such that the map

$$\mathcal{P}^{(n)} \to \mathcal{P}_{\mathcal{S}}^{(n)}$$
$$g \mapsto \pi(g)$$

is bijective, where $\pi \colon \mathcal{G} \to \mathcal{S}$ is the natural surjective homomorphism.

Remark 2.68. For all $P, Q \in L(\mathcal{S})$ it holds

$$(P, \tilde{\tau}(P))(Q, \tilde{\tau}(Q)) = (I_{d_2}, \bar{\tau}(P, Q))(PQ, \tilde{\tau}(PQ))$$

and thus, the map $\bar{\tau}$ is well-defined.

If $n = 1$, then $a(n) = 0$ and $\mathcal{P}_{\mathcal{S}}^{(n)} = \{(P, \tilde{\tau}(P)) \mid P \in L(\mathcal{S})\}$.

Lemma 2.69. *For all $n \in \mathbb{N}$ coprime to $|L(\mathcal{S})|$ and for all $N \in (n\mathbb{N}) \cap M_0$
it holds*
$$\mathcal{T}^n \mathcal{F} \mathcal{P}^{(n)} \leq \mathcal{G} \quad \text{and} \quad \mathcal{T}^N \triangleleft \mathcal{T}^n \mathcal{F} \mathcal{P}^{(n)}.$$

Proof. Let $n \in \mathbb{N}$ be coprime to $|L(\mathcal{S})|$.

First, we prove that $\mathcal{T}_{\mathcal{S}}^n \mathcal{P}_{\mathcal{S}}^{(n)}$ is a subgroup of $\mathcal{S}$. Let $t, s \in \mathcal{T}_{\mathcal{S}}^n$ and
$p, q \in \mathcal{P}_{\mathcal{S}}^{(n)}$. We have to show that $tp(sq)^{-1} \in \mathcal{T}_{\mathcal{S}}^n \mathcal{P}_{\mathcal{S}}^{(n)}$. Clearly, it
holds $tp(sq)^{-1} = tpq^{-1}s^{-1}(pq^{-1})^{-1}pq^{-1}$. Since $\mathcal{T}_{\mathcal{S}}^n \triangleleft \mathcal{S}$, we have that
$(pq^{-1})s^{-1}(pq^{-1})^{-1} \in \mathcal{T}_{\mathcal{S}}^n$, and hence, it suffices to show that $pq^{-1} \in$
$\mathcal{T}_{\mathcal{S}}^n \mathcal{P}_{\mathcal{S}}^{(n)}$. Let $P = L(p)$, $Q = L(q)$ and $R = PQ^{-1} \in L(\mathcal{S})$. Let $a = \max\{\tilde{a} \in \{0, -1, \dots\} \mid \exists b \in \mathbb{Z} \text{ such that } a|L(\mathcal{S})| + bn = 1\}$ and $b \in \mathbb{Z}$
such that $a|L(\mathcal{S})| + bn = 1$. We compute

$$pq^{-1} = \left(P, \tilde{\tau}(P) - a \sum_{S \in L(\mathcal{S})} \bar{\tau}(P, S)\right)\left(Q^{-1}, -Q^{-1}\tilde{\tau}(Q)\right.$$

$$\left. + a \sum_{S \in L(\mathcal{S})} Q^{-1}\bar{\tau}(Q, S)\right)$$

$$= \left(R, \tilde{\tau}(P) - PQ^{-1}\tilde{\tau}(Q) - a \sum_{S \in L(\mathcal{S})} (\bar{\tau}(P, S) - PQ^{-1}\bar{\tau}(Q, S))\right)$$

$$= \left(R, \tilde{\tau}(R) - \bar{\tau}(PQ^{-1}, Q) - a \sum_{S \in L(\mathcal{S})} (\bar{\tau}(P, S) - PQ^{-1}\bar{\tau}(Q, S))\right)$$

$$= \left(R, \tilde{\tau}(R) - (a|L(\mathcal{S})| + bn)\bar{\tau}(PQ^{-1}, Q)\right.$$

$$\left. - a \sum_{S \in L(\mathcal{S})} (\bar{\tau}(P, S) - PQ^{-1}\bar{\tau}(Q, S))\right)$$

$$= \left(I_{d_2}, \bar{\tau}(R,Q)\right)^{-bn} \Bigg(R, \tilde{\tau}(R)$$

$$- a \sum_{S \in L(\mathcal{S})} (\bar{\tau}(PQ^{-1}, Q) + \bar{\tau}(P,S) - PQ^{-1}\bar{\tau}(Q,S))\Bigg)$$

$$= \left(I_{d_2}, \bar{\tau}(R,Q)\right)^{-bn} \Bigg(R, \tilde{\tau}(R)$$

$$- a \sum_{S \in L(\mathcal{S})} (\tilde{\tau}(PQ^{-1}) - \tilde{\tau}(PS) + PQ^{-1}\tilde{\tau}(QS))\Bigg).$$

We use that $\sum_{S \in L(\mathcal{S})} \tilde{\tau}(S) = \sum_{S \in L(\mathcal{S})} \tilde{\tau}(TS)$ for all $T \in L(\mathcal{S})$.

$$pq^{-1} = \left(I_{d_2}, \bar{\tau}(R,Q)\right)^{-bn} \Bigg(R, \tilde{\tau}(R)$$

$$- a \sum_{S \in L(\mathcal{S})} (\tilde{\tau}(PQ^{-1}) - \tilde{\tau}(PQ^{-1}S) + PQ^{-1}\tilde{\tau}(S))\Bigg)$$

$$= \left(I_{d_2}, \bar{\tau}(R,Q)\right)^{-bn} \Bigg(R, \tilde{\tau}(R) - a \sum_{S \in L(\mathcal{S})} \bar{\tau}(R,S)\Bigg) \in \mathcal{T}_\mathcal{S}^n \mathcal{P}_\mathcal{S}^{(n)}.$$

Thus, we have $\mathcal{T}_\mathcal{S}^n \mathcal{P}_\mathcal{S}^{(n)} \leq \mathcal{S}$.

Let π be the natural surjective homomorphism from $\mathcal{G}$ to $\mathcal{S}$ with kernel $\mathcal{F}$. It holds $\pi^{-1}(\mathcal{T}_\mathcal{S}^n \mathcal{P}_\mathcal{S}^{(n)}) = T^n \mathcal{F} \mathcal{P}^{(n)}$ and thus, $T^n \mathcal{F} \mathcal{P}^{(n)}$ is a subgroup of $\mathcal{G}$.

Now let $N \in (n\mathbb{N}) \cap M_0$. Since n divides N, we have $T^N \subset T^n \mathcal{F} \mathcal{P}^{(n)}$. Since $N \in M_0$, we have $T^N \triangleleft T^n \mathcal{F} \mathcal{P}^{(n)}$. □

Recall Definition 2.35.

Remark 2.70. Let $n \in \mathbb{N}$ be coprime to $|L(\mathcal{S})|$. Let $m \in M_0$, $N = nm$ and $t_1, \ldots, t_{d_2} \in T^n$ such that $\pi(\{t_1, \ldots, t_{d_2}\})$ generates $\mathcal{T}_\mathcal{S}^n$, where $\pi \colon T\mathcal{F} \to \mathcal{T}_\mathcal{S}$ is the natural surjective homomorphism. Then, the map

$$\{0, \ldots, m-1\}^{d_2} \times \mathcal{F} \times \mathcal{P}^{(n)} \to (T^n \mathcal{F} \mathcal{P}^{(n)})_N$$

$$((n_1, \ldots, n_{d_2}), f, p) \mapsto t_1^{n_1} \ldots t_{d_2}^{n_{d_2}} fp T^N$$

is bijective.

The following lemma characterizes the elements of the finite groups $\mathcal{G}_N$, $(T^n \mathcal{F})_N$ and $(T^m)_N$ for appropriate $n, m, N \in \mathbb{N}$.

Lemma 2.71. *Let $t_1, \ldots, t_{d_2} \in \mathcal{T}$ such that the set $\pi(\{t_1, \ldots, t_{d_2}\})$ generates $\mathcal{T}_S$, where $\pi \colon \mathcal{T}\mathcal{F} \to \mathcal{T}_S$ is the natural surjective homomorphism. For all $N \in M_0$ it holds*

$$\mathcal{G}_N = \left\{ t_1^{n_1} \ldots t_{d_2}^{n_{d_2}} f p \mathcal{T}^N \,\middle|\, n_1, \ldots, n_{d_2} \in \{0, \ldots, N-1\}, f \in \mathcal{F}, p \in \mathcal{P}^{(1)} \right\}$$

and particularly $|\mathcal{G}_N| = N^{d_2} |\mathcal{F}| |L(\mathcal{S})|$.
For all $n \in \mathbb{N}$ and $N \in (n\mathbb{N}) \cap M_0$ it holds

$$(\mathcal{T}^n \mathcal{F})_N = \left\{ t_1^{n n_1} \ldots t_{d_2}^{n n_{d_2}} f \mathcal{T}^N \,\middle|\, n_1, \ldots, n_{d_2} \in \{0, \ldots, N/n - 1\}, f \in \mathcal{F} \right\}$$

and particularly $|(\mathcal{T}^n \mathcal{F})_N| = (N/n)^{d_2} |\mathcal{F}|$. Moreover, for all $n \in \mathbb{N}$ and $N \in (n\mathbb{N}) \cap M_0$ it holds $(\mathcal{T}^n \mathcal{F})_N \triangleleft \mathcal{G}_N$.
For all $m \in M_0$ and $N \in m\mathbb{N}$ it holds

$$(\mathcal{T}^m)_N = \left\{ t_1^{m n_1} \ldots t_{d_2}^{m n_{d_2}} \mathcal{T}^N \,\middle|\, n_1, \ldots, n_{d_2} \in \{0, \ldots, N/m - 1\} \right\},$$

$(\mathcal{T}^m)_N$ is a subgroup of the center of $(\mathcal{T}\mathcal{F})_N$ and particularly $|(\mathcal{T}^m)_N| = (N/m)^{d_2}$.

Proof. Since $\mathcal{P}^{(1)}$ is a representation set of $\mathcal{G}/\mathcal{T}\mathcal{F}$, the map $\mathcal{T} \times \mathcal{F} \times \mathcal{P}^{(1)} \to \mathcal{G}$, $(t, f, p) \mapsto tfp$ is bijective. The assertions are clear by Lemma 2.11, Theorem 2.17 and Lemma 2.12, Proposition 2.15. $\square$

The following theorem characterizes the group $\mathcal{G}_N$ for appropriate $N \in \mathbb{N}$.

Theorem 2.72. *Let $m \in M_0$. Let $n \in \mathbb{N}$ be coprime to m and $|L(\mathcal{S})|$. Let $N = nm$. Then, we have*

$$\mathcal{G}_N = (\mathcal{T}^m)_N \rtimes (\mathcal{T}^n \mathcal{F} \mathcal{P}^{(n)})_N$$

and $(\mathcal{T}^m)_N$ is isomorphic to $\mathbb{Z}_n^{d_2}$.

Proof. Let $m \in M_0$. Let $n \in \mathbb{N}$ be coprime to m and $|L(\mathcal{S})|$. Let $N = nm$. By Theorem 2.17 we have $\mathcal{T}^m \triangleleft \mathcal{G}$ and $\mathcal{T}^N \triangleleft \mathcal{G}$, and by Lemma 2.12 we have $\mathcal{T}^N \triangleleft \mathcal{T}^m$. Hence, we have

$$(\mathcal{T}^m)_N \triangleleft \mathcal{G}_N. \tag{2.9}$$

By Lemma 2.12 the group $\mathcal{T}^m$ is isomorphic to $\mathbb{Z}^{d_2}$ and thus, $(\mathcal{T}^m)_N$ is isomorphic to $\mathbb{Z}_n^{d_2}$. By Lemma 2.69 we have

$$(\mathcal{T}^n \mathcal{F} \mathcal{P}^{(n)})_N \le \mathcal{G}_N. \tag{2.10}$$

For all $N \in \mathbb{N}$ and $\mathcal{H} \leq \mathcal{S}$ such that $\mathcal{T}_\mathcal{S}^N$ is a subgroup of $\mathcal{H}$, we denote

$$\mathcal{H}_N := \mathcal{H}/\mathcal{T}_\mathcal{S}^N.$$

Let $\pi \colon \mathcal{G}_N \to \mathcal{S}_N$ be the natural surjective homomorphism with kernel $\{g\mathcal{T}^N \mid g \in \mathcal{F}\}$. We have

$$\pi((\mathcal{T}^m)_N \cap (\mathcal{T}^n \mathcal{F} \mathcal{P}^{(n)})_N) \subset \pi((\mathcal{T}^m)_N) \cap \pi((\mathcal{T}^n \mathcal{F} \mathcal{P}^{(n)})_N)$$
$$= (\mathcal{T}_\mathcal{S}^m)_N \cap (\mathcal{T}_\mathcal{S}^n \mathcal{P}_\mathcal{S}^{(n)})_N$$
$$= (\mathcal{T}_\mathcal{S}^m)_N \cap (\mathcal{T}_\mathcal{S}^n)_N$$
$$= \{id\}, \tag{2.11}$$

where in the third step we used that for all $p \in \mathcal{P}_\mathcal{S}^{(n)}$ such that $L(p) = I_{d_2}$ we have $p \in \mathcal{T}_\mathcal{S}^n$ and in the last step we used that the numbers n^{d_2} and m^{d_2} are coprime, $|(\mathcal{T}_\mathcal{S}^m)_N| = n^{d_2}$, $|(\mathcal{T}_\mathcal{S}^n)_N| = m^{d_2}$ and Lagrange's theorem. By (2.11) and since $\pi|_{(\mathcal{T}^m)_N}$ is injective, we have

$$(\mathcal{T}^m)_N \cap (\mathcal{T}^n \mathcal{F} \mathcal{P}^{(n)})_N = \{id\}. \tag{2.12}$$

We have

$$|\mathcal{G}_N| = |\ker(\pi)||\pi(\mathcal{G}_N)| = |\{g\mathcal{T}^N \mid g \in \mathcal{F}\}||\mathcal{S}_N| = |\mathcal{F}||L(\mathcal{S})|N^{d_2}, \tag{2.13}$$

see Lemma 2.71, and

$$|(\mathcal{T}^n \mathcal{F} \mathcal{P}^{(n)})_N| = |\ker(\pi|_{(\mathcal{T}^n \mathcal{F} \mathcal{P}^{(n)})_N})||\pi((\mathcal{T}^n \mathcal{F} \mathcal{P}^{(n)})_N)|$$
$$= |\mathcal{F}||(\mathcal{T}_\mathcal{S}^n \mathcal{P}_\mathcal{S}^{(n)})_N| = |\mathcal{F}||\mathcal{P}_\mathcal{S}^{(n)}||(\mathcal{T}_\mathcal{S}^n)_N| = |\mathcal{F}||L(\mathcal{S})|m^{d_2}, \tag{2.14}$$

see Remark 2.70. By (2.13), (2.14) and since $(\mathcal{T}^m)_N$ is isomorphic to $\mathbb{Z}_n^{d_2}$, we have

$$|\mathcal{G}_N| = |(\mathcal{T}^m)_N||(\mathcal{T}^n \mathcal{F} \mathcal{P}^{(n)})_N|. \tag{2.15}$$

By (2.9), (2.10), (2.12) and (2.15) we have

$$\mathcal{G}_N = (\mathcal{T}^m)_N \rtimes (\mathcal{T}^n \mathcal{F} \mathcal{P}^{(n)})_N. \qquad \square$$

Corollary 2.73. *Let* $m \in M_0$, $\tilde{n} \in \mathbb{N}$, $n = \tilde{n}m|L(\mathcal{S})| + 1$ *and* $N = nm$. *Then we have*

$$\mathcal{P}_\mathcal{S}^{(n)} = \left\{ \left(P, \tilde{\tau}(P) + \tilde{n}m \sum_{Q \in L(\mathcal{S})} \bar{\tau}(P,Q) \right) \,\middle|\, P \in L(\mathcal{S}) \right\}$$

and

$$\mathcal{G}_N = (\mathcal{T}^m)_N \rtimes (\mathcal{T}^n \mathcal{F} \mathcal{P}^{(n)})_N.$$

Proof. Let $m \in M_0$, $\tilde{n} \in \mathbb{N}$, $n = \tilde{n}m|L(\mathcal{S})|+1$ and $N = nm$. In particular, n is coprime to m and $|L(\mathcal{S})|$. We have

$$\max\{\tilde{a} \in \{0, -1, \ldots\} \mid \exists b \in \mathbb{Z} \text{ such that } \tilde{a}|L(\mathcal{S})| + bn = 1\}$$
$$= \max\{\tilde{a} \in \{0, -1, \ldots\} \mid \exists b \in \mathbb{N} \text{ such that } (\tilde{a} + b\tilde{n}m)|L(\mathcal{S})| + b = 1\}$$
$$= -\tilde{n}m$$

and hence,

$$\mathcal{P}_\mathcal{S}^{(n)} = \left\{ \left(P, \tilde{\tau}(P) + \tilde{n}m \sum_{Q \in L(\mathcal{S})} \bar{\tau}(P, Q) \right) \,\Big|\, P \in L(\mathcal{S}) \right\}.$$

By Theorem 2.72 we have $\mathcal{G}_N = (\mathcal{T}^m)_N \rtimes (\mathcal{T}^n \mathcal{F} \mathcal{P}^{(n)})_N$. □

Corollary 2.74. *Suppose that $\mathcal{G}$ is a space group. Let $N \in \mathbb{N}$ be coprime to $|L(\mathcal{G})|$. Then we have*

$$\mathcal{G}_N = \mathcal{T}_N \rtimes \{g\mathcal{T}^N \mid g \in \mathcal{P}^{(N)}\}.$$

Proof. Let $\mathcal{G}$ be a space group. We have $\mathcal{F} = \{id\}$ and $M_0 = \mathbb{N}$. For all $N \in \mathbb{N}$ coprime to $|L(\mathcal{G})|$, we have $(\mathcal{T}^N \mathcal{P}^{(N)})/\mathcal{T}^N = \{g\mathcal{T}^N \mid g \in \mathcal{P}^{(N)}\}$. Thus, Theorem 2.72 implies the assertion. □

Corollary 2.75. *Suppose that $\mathcal{G}$ is a space group. Let $n \in \mathbb{N}$ and $N = n|L(\mathcal{G})| + 1$. Then it holds*

$$\mathcal{P}^{(N)} = \left\{ \left(P, \tilde{\tau}(P) + n \sum_{Q \in L(\mathcal{G})} \bar{\tau}(P, Q) \right) \,\Big|\, P \in L(\mathcal{G}) \right\}$$

and

$$\mathcal{G}_N = \mathcal{T}_N \rtimes \{g\mathcal{T}^N \mid g \in \mathcal{P}^{(N)}\}.$$

Proof. This is clear by Corollary 2.73 and Corollary 2.74. □

Corollary 2.76. *Suppose that $\mathcal{G} = \mathcal{T}\mathcal{F}$. Let $m \in M_0$ and $n \in \mathbb{N}$ be coprime. Let $N = nm$. Then it holds*

$$\mathcal{G}_N = (\mathcal{T}^m)_N \times (\mathcal{T}^n \mathcal{F})_N.$$

Proof. Suppose that $\mathcal{G} = \mathcal{T}\mathcal{F}$. Let $m \in M_0$ and $n \in \mathbb{N}$ be coprime. We have $\mathcal{S} = \mathcal{T}_\mathcal{S}$ and $L(\mathcal{S}) = \{I_{d_2}\}$. Without loss of generality we assume that $\tilde{\tau} = 0$. We have $\bar{\tau} = 0$ and $\mathcal{P}_\mathcal{S}^{(n)} = \{id\}$. Without loss of generality we assume that $\mathcal{P}^{(n)} = \{id\}$. By Theorem 2.72, Lemma 2.11(ii) and Proposition 2.15 we have $\mathcal{G}_N = (\mathcal{T}^m)_N \times (\mathcal{T}^n \mathcal{F})_N$. □

2.4. Orbits of discrete subgroups of the Euclidean group

James [41] defined an objective atomic structure which is an orbit of a point under the action of a discrete subgroup of $E(d)$, see [42, Proposition 3.14]. In this thesis we consider only orbits where the stabilizer subgroup is trivial and thus we have a natural bijection between the discrete group and the atoms.

Definition 2.77. We call a subset S of $\mathbb{R}^d$ a *general configuration* if there exist a discrete group $\mathcal{G} < E(d)$ and a point $x \in \mathbb{R}^d$ such that the map $\mathcal{G} \to S$, $g \mapsto g \cdot x$ is bijective.

Remark 2.78. (i) For each discrete group $\mathcal{G} < E(d)$ there exists a point $x \in \mathbb{R}^d$ such that the map $\mathcal{G} \to \mathbb{R}^d$, $g \mapsto g \cdot x$ is injective, see, e.g., [18, Appendix A.3]. In particular, the set $\mathcal{G} \cdot x$ is a general configuration.

(ii) The representation of a general configuration by a discrete subgroup of $E(d)$ and a point in $\mathbb{R}^d$ is not unique, see Example 2.79. Moreover, the orbit of a point in $\mathbb{R}^d$ under the action of a discrete subgroup of $E(d)$ need not be a general configuration, see Example 2.80.

Example 2.79. We present an example showing that in general for a given general configuration $S \subset \mathbb{R}^d$ there exist discrete groups $\mathcal{G}_1, \mathcal{G}_2 < E(d)$ and a point $x \in \mathbb{R}^d$ such that the maps $\mathcal{G}_1 \to S$, $g \mapsto g \cdot x$ and $\mathcal{G}_2 \to S$, $g \mapsto g \cdot x$ are bijective but $\mathcal{G}_1$ and $\mathcal{G}_2$ are not isomorphic. Let $S = \{\pm e_1, \pm e_2\} \subset \mathbb{R}^2$, $\mathcal{G}_1 = \langle (R(\pi/2), 0) \rangle < E(2)$,

$$\mathcal{G}_2 = \left\langle \left(\begin{pmatrix} 0 & 1 \\ 1 & 0 \end{pmatrix}, 0 \right), \left(\begin{pmatrix} 0 & -1 \\ -1 & 0 \end{pmatrix}, 0 \right) \right\rangle < E(2)$$

and $x = e_1 \in \mathbb{R}^2$. The group $\mathcal{G}_2$ is the Klein four-group and thus, $\mathcal{G}_1$ and $\mathcal{G}_2$ are not isomorphic. However, the maps $\mathcal{G}_1 \to S$, $g \mapsto g \cdot x$ and $\mathcal{G}_2 \to S$, $g \mapsto g \cdot x$ are bijective.

Example 2.80. In this example we present an orbit S of a point in $\mathbb{R}^3$ under the action of a discrete subgroup of $E(3)$ which is not a general configuration.

Let be given a regular icosahedron centered at the origin. Let S be the set of the 30 centers of the edges of the icosahedron (i.e. S is the set of the vertices of the rectified icosahedron and moreover, S is the set of the vertices of a icosidodecahedron). The rotation group $\mathcal{I} < \mathrm{SO}(3)$ of the icosahedron has order 60, see, e. g., [36, Section 2.4] and we have $S = (\mathcal{I} \times \{0_3\}) \cdot x_0$ for every point $x_0 \in S$. Now we suppose that there exist a discrete group $\mathcal{G} < \mathrm{E}(3)$ and a point $x \in \mathbb{R}^3$ such that the map $\mathcal{G} \to S$, $g \mapsto g \cdot x$ is injective. Then we have $|\mathcal{G}| = |S| = 30$. Moreover, the group $\mathcal{G}$ is isomorphic to a finite subgroup of $\mathrm{O}(3)$, see, e. g., [47, Section 4.12]. The finite subgroups of $\mathrm{O}(3)$ are classified, see, e. g., [36, Theorem 2.5.2], and since every discrete subgroup of $\mathrm{O}(3)$ of order 30 contains an element of order 15, the group $\mathcal{G}$ contains an element g of order 15. Since the order of g is odd, we have $L(g) \in \mathrm{SO}(3)$, i. e. g is a rotation. Thus, the set S contains 15 points which lie in the same plane. This implies that S cannot be the orbit of $\mathcal{G}$, and we have a contradiction.

Lemma 2.81. *Let $S \subset \mathbb{R}^d$ be a general configuration. Then for all $a \in \mathrm{E}(d)$ the set $\{a \cdot x \,|\, x \in S\}$ is also a general configuration.*

Proof. Let $S \subset \mathbb{R}^d$ be a general configuration. There exist some discrete group $\mathcal{G} < \mathrm{E}(d)$ and $x_0 \in \mathbb{R}^d$ such that the map

$$\mathcal{G} \to S$$
$$g \mapsto g \cdot x_0$$

is bijective. Then, for every $a \in \mathrm{E}(d)$ the map

$$a\mathcal{G}a^{-1} \to \{a \cdot x \,|\, x \in S\}$$
$$g \mapsto g \cdot (a \cdot x_0)$$

is bijective and thus, the set $\{a \cdot x \,|\, x \in S\}$ is a general configuration. □

The following definition can be found in, e. g., [59, p. 14].

Definition 2.82. For all $A \subset \mathbb{R}^n$ we define

$$\dim(A) := \dim(\mathrm{aff}(A)),$$

where $\mathrm{aff}(A)$ is the affine hull of A.

The following lemma is clear by the above definition.

Lemma 2.83. *For all $A \subset \mathbb{R}^d$ and $x_0 \in A$ it holds*

$$\dim(A) = \dim(\mathrm{span}(\{x - x_0 \,|\, x \in A\})).$$

Lemma 2.84. *Let $\mathcal{G} < \mathrm{E}(d)$ be discrete and $x_0 \in \mathbb{R}^d$ such that the map $\mathcal{G} \to \mathbb{R}^d$, $g \mapsto g \cdot x_0$ is injective. Let $d_{\mathrm{aff}} = \dim(\mathcal{G} \cdot x_0)$. Then there exists some $a \in \mathrm{E}(d)$ such that for the discrete group $\mathcal{G}' = a\mathcal{G}a^{-1}$ and $x_0' = a \cdot x_0$ it holds*

$$\mathrm{aff}(\mathcal{G}' \cdot x_0') = \{0_{d-d_{\mathrm{aff}}}\} \times \mathbb{R}^{d_{\mathrm{aff}}}.$$

The map $\mathcal{G}' \to \mathbb{R}^d$, $g \mapsto g \cdot x_0'$ is injective and we have $\mathcal{G}' \cdot x_0' = a \cdot (\mathcal{G} \cdot x_0)$.

Proof. Let $\mathcal{G} < \mathrm{E}(d)$ be discrete and $x_0 \in \mathbb{R}^d$ such that the map $\mathcal{G} \to \mathbb{R}^d$, $g \mapsto g \cdot x_0$ is injective. Let $d_{\mathrm{aff}} = \dim(\mathcal{G} \cdot x_0)$. There exists some d_{aff}-dimensional vector space V such that $\mathrm{aff}(\mathcal{G} \cdot x_0) = x_0 + V$. There exists some $A \in \mathrm{O}(d)$ such that $\{Ax \,|\, x \in V\} = \{0_{d-d_{\mathrm{aff}}}\} \times \mathbb{R}^{d_{\mathrm{aff}}}$. The choice $a = (A, -Ax_0) \in \mathrm{E}(d)$ implies the assertion. $\square$

Lemma 2.85. *Let $\mathcal{G} < \mathrm{E}(d)$ be discrete and $x_0 \in \mathbb{R}^d$ such that $\mathrm{aff}(\mathcal{G} \cdot x_0) = \{0_{d-d_{\mathrm{aff}}}\} \times \mathbb{R}^{d_{\mathrm{aff}}}$, where $d_{\mathrm{aff}} = \dim(\mathrm{aff}(\mathcal{G} \cdot x_0))$. Then we have $\mathcal{G} < \mathrm{O}(d - d_{\mathrm{aff}}) \oplus \mathrm{E}(d_{\mathrm{aff}})$.*

Proof. Let $\mathcal{G} < \mathrm{E}(d)$ be discrete and $x_0 \in \mathbb{R}^d$ such that $V = \{0_{d-d_{\mathrm{aff}}}\} \times \mathbb{R}^{d_{\mathrm{aff}}}$, where $V = \mathrm{aff}(\mathcal{G} \cdot x_0)$ and $d_{\mathrm{aff}} = \dim(\mathrm{aff}(\mathcal{G} \cdot x_0))$. Let $g \in \mathcal{G}$. We define the map $\varphi \colon \mathbb{R}^d \to \mathbb{R}^d$, $x \mapsto L(g)x$.
First we show that V is invariant under φ. Let $x \in V$. Since $V = \mathrm{aff}(\mathcal{G} \cdot x_0) - x_0$, there exist some $n \in \mathbb{N}$, $x_1, \ldots, x_n \in \mathcal{G} \cdot x_0$ and $\alpha_1, \ldots, \alpha_n \in \mathbb{R}$ such that $x = \sum_{i=1}^{n} \alpha_i x_i$ and $\sum_{i=1}^{n} \alpha_i = 0$. It holds

$$L(g)x = \sum_{i=1}^{n} \alpha_i L(g)x_i = \sum_{i=1}^{n} \alpha_i (g \cdot x_i) \in V.$$

Thus we have $\{L(g)\tilde{x} \,|\, \tilde{x} \in V\} \subset V$. Since $L(g)$ is invertible, it holds $\{L(g)\tilde{x} \,|\, \tilde{x} \in V\} = V$.
Since $L(g)$ is orthogonal, also the complement $V^{\perp} = \mathbb{R}^{d-d_{\mathrm{aff}}} \times \{0_{d_{\mathrm{aff}}}\}$ is invariant under φ. This implies $L(g) \in \mathrm{O}(d_1) \oplus \mathrm{O}(d_2)$. It holds $\tau(g) = g \cdot x_0 - L(g)x_0 \in V$ and thus, $g \in \mathrm{O}(d_1) \oplus \mathrm{E}(d_2)$. $\square$

Lemma 2.86. *Let $\mathcal{G} < \mathrm{E}(d)$ be discrete and $x_0 \in \mathbb{R}^d$ such that the map $\mathcal{G} \to \mathbb{R}^d$, $g \mapsto g \cdot x_0$ is injective and $\mathrm{aff}(\mathcal{G} \cdot x_0) = \{0_{d-d_{\mathrm{aff}}}\} \times \mathbb{R}^{d_{\mathrm{aff}}}$, where $d_{\mathrm{aff}} = \dim(\mathcal{G} \cdot x_0)$. Let $\mathcal{G}' = \{I_{d-d_{\mathrm{aff}}} \oplus g \,|\, g \in \mathrm{E}(d_2), \exists A \in \mathrm{O}(d_1) : A \oplus g \in \mathcal{G}\}$.*
Then $\mathcal{G}'$ is a discrete subgroup of $\mathrm{E}(d)$, the map

$$\mathcal{G} \to \mathcal{G}'$$
$$A \oplus g \mapsto I_{d-d_{\mathrm{aff}}} \oplus g \qquad \textit{if } A \in \mathrm{O}(d - d_{\mathrm{aff}}), g \in \mathrm{E}(d_{\mathrm{aff}}) \textit{ and } A \oplus g \in \mathcal{G}$$

is an isomorphism, $\mathcal{G} \cdot x_0 = \mathcal{G}' \cdot x_0$ and the map $\mathcal{G}' \to \mathbb{R}^d$, $g \mapsto g \cdot x_0$ is injective.

Proof. Let $\mathcal{G} < \mathrm{E}(d)$ be discrete and $x_0 \in \mathbb{R}^d$ such that the map $\mathcal{G} \to \mathbb{R}^d$, $g \mapsto g \cdot x_0$ is injective and $\mathrm{aff}(\mathcal{G} \cdot x_0) = \{0_{d-d_{\mathrm{aff}}}\} \times \mathbb{R}^{d_{\mathrm{aff}}}$, where $d_{\mathrm{aff}} = \dim(\mathcal{G} \cdot x_0)$. By Lemma 2.85 we have $\mathcal{G} < \mathrm{O}(d - d_{\mathrm{aff}}) \oplus \mathrm{E}(d_{\mathrm{aff}})$. Let $\mathcal{G}' = \{I_{d-d_{\mathrm{aff}}} \oplus g \mid g \in \mathrm{E}(d_2), \exists A \in \mathrm{O}(d_1) : A \oplus g \in \mathcal{G}\}$. We define the map

$$\varphi \colon \mathcal{G} \to \mathcal{G}'$$
$$A \oplus g \mapsto I_{d-d_{\mathrm{aff}}} \oplus g \quad \text{if } A \in \mathrm{O}(d - d_{\mathrm{aff}}), g \in \mathrm{E}(d_{\mathrm{aff}}) \text{ and } A \oplus g \in \mathcal{G}.$$

It is clear that φ is a surjective homomorphism. Since $x_0 \in \{0_{d-d_{\mathrm{aff}}}\} \times \mathbb{R}^{d_{\mathrm{aff}}}$, for all $g \in \mathcal{G}$ it holds $g \cdot x_0 = \varphi(g) \cdot x_0$. Particularly, we have $\mathcal{G} \cdot x_0 = \mathcal{G}' \cdot x_0$. Since the map $\mathcal{G} \to \mathbb{R}^d$, $g \mapsto g \cdot x_0$ is injective, the map φ is injective and thus, an isomorphism. Since the map $\mathcal{G}' \to \mathcal{G}' \cdot x_0$, $g \mapsto g \cdot x_0$ is a homeomorphism and $\mathcal{G}' \cdot x_0$ is discrete, $\mathcal{G}'$ is discrete. $\square$

Remark 2.87.   (i) Let $\mathcal{G} < \mathrm{E}(d)$ be discrete, $x_0 \in \mathbb{R}^d$ and $A = \mathrm{aff}(\mathcal{G} \cdot x_0)$. For all $g \in \mathcal{G}$ it holds $\{g \cdot x \mid x \in A\} = A$.

(ii) Let $\mathcal{G} < \mathrm{E}(d)$ be discrete and $x_0 \in \mathbb{R}^d$. Let V be the vector space such that $\mathrm{aff}(\mathcal{G} \cdot x_0) = x_0 + V$. Then for all $g \in \mathcal{G}$ it holds $\{L(g)x \mid x \in V\} = V$.

3. Seminorms on the vector space of all periodic displacements

The main results of this chapter are Theorem 3.34, Theorem 3.37 and Theorem 3.40.

We use the following notation. Let d, d_1, d_2, $\mathcal{G}$ and $\mathcal{T}$ be as in Definition 2.6, M_0 as in Definition 2.13 and $\mathcal{C}_N$ as in Definition 2.50 for all $N \in M_0$. Let $x_0 \in \mathbb{R}^d$ be such that the map $\mathcal{G} \to \mathbb{R}^d$, $g \mapsto g \cdot x_0$ is injective. Let d_{aff} denote the dimension $\dim(\mathcal{G} \cdot x_0)$. Moreover we suppose that

$$\text{aff}(\mathcal{G} \cdot x_0) = \{0_{d-d_{\text{aff}}}\} \times \mathbb{R}^{d_{\text{aff}}},$$

which can be achieved by a coordinate transformation, see Lemma 2.84.

3.1. Motivation of the model and the seminorms

In the physical model the dimension d is equal to 3 and there are atoms at the points $\mathcal{G} \cdot x_0$. Since the map $\mathcal{G} \to \mathbb{R}^d$, $g \mapsto g \cdot x_0$ is injective, we have a canonical bijection between $\mathcal{G}$ and the atoms. We displace the atoms a little bit and describe the small displacement by a function $u \colon \mathcal{G} \to \mathbb{R}^d$ such that the atoms are now at the points

$$(v_u(g))_{g \in \mathcal{G}} := (g \cdot (x_0 + u(g)))_{g \in \mathcal{G}}.$$

If $u = 0$, then the atoms are not displaced. If there exists some $a \in \mathbb{R}^d$ such that $L(g)u(g) = a$ for all $g \in \mathcal{G}$, then we have a translation of the atoms in the physical model. If there exists some $R \in \text{SO}(d)$ such that $L(g)u(g) = (R - I_d)(g \cdot x_0)$ for all $g \in \mathcal{G}$, then we have a rotation of the atoms about the origin in the physical model.

Let $\mathcal{R}$ be a finite appropriate subset of $\mathcal{G}$. Now we want to define a seminorm $\| \cdot \|_{\mathcal{R}}$ on the vector space of all appropriate displacements u, which quantifies the size of v_u 'modulo local isometries'. Let $u \colon \mathcal{G} \to \mathbb{R}^d$ be $\mathcal{T}^N$-periodic for some $N \in M_0$. We want to define the seminorm such that

$$\|u\|_{\mathcal{R}} \approx \left(\frac{1}{|\mathcal{C}_N|} \sum_{g \in \mathcal{C}_N} \operatorname{dist}^2 \left(v_u|_{g\mathcal{R}}, \left\{ (a \cdot (h \cdot x_0))_{h \in g\mathcal{R}} \,\middle|\, a \in \mathrm{E}(d) \right\} \right) \right)^{\frac{1}{2}},$$

$$(3.1)$$

where dist is the induced metric of the Euclidean norm on $(\mathbb{R}^d)^{\mathcal{R}}$. For every $g \in \mathcal{C}_N$ we have

$$\operatorname{dist}\left(v_u|_{g\mathcal{R}}, \left\{ (a \cdot (h \cdot x_0))_{h \in g\mathcal{R}} \,\middle|\, a \in \mathrm{E}(d) \right\} \right)$$

$$= \operatorname{dist}\left(\left((gh) \cdot (x_0 + u(gh)) \right)_{h \in \mathcal{R}}, \left\{ ((ah) \cdot x_0)_{h \in \mathcal{R}} \,\middle|\, a \in \mathrm{E}(d) \right\} \right)$$

$$= \operatorname{dist}\left((u(gh))_{h \in \mathcal{R}}, \left\{ \left(L(h)^\mathsf{T}((ah) \cdot x_0 - h \cdot x_0) \right)_{h \in \mathcal{R}} \,\middle|\, a \in \mathrm{E}(d) \right\} \right).$$

$$(3.2)$$

Let $U \subset \mathrm{E}(d)$ be a sufficiently small open neighborhood of id. Then the set

$$\left\{ \left(L(h)^\mathsf{T}((ah) \cdot x_0 - h \cdot x_0) \right)_{h \in \mathcal{R}} \,\middle|\, a \in U \right\}$$

is a manifold and its tangent space at the point $0 \in (\mathbb{R}^d)^{\mathcal{R}}$ is

$$U_{\mathrm{iso}}(\mathcal{R}) = \left\{ \left(L(h)^\mathsf{T}(b + S(h \cdot x_0)) \right)_{h \in \mathcal{R}} \,\middle|\, b \in \mathbb{R}^d, S \in \mathrm{Skew}(d) \right\},$$

see Proposition 4.19. Since we consider only small displacements $u \approx 0$, by (3.2) and Taylor's theorem we have

$$\operatorname{dist}\left(v_u|_{g\mathcal{R}}, \left\{ (a \cdot (h \cdot x_0))_{h \in g\mathcal{R}} \,\middle|\, a \in \mathrm{E}(d) \right\} \right) \approx \operatorname{dist}(u(g \cdot)|_{\mathcal{R}}, U_{\mathrm{iso}}(\mathcal{R})).$$

$$(3.3)$$

By (3.1) and (3.3) we define the seminorm $\|\cdot\|_{\mathcal{R}}$ by

$$\|u\|_{\mathcal{R}} = \left(\frac{1}{|\mathcal{C}_N|} \sum_{g \in \mathcal{C}_N} \operatorname{dist}^2 (u(g \cdot)|_{\mathcal{R}}, U_{\mathrm{iso}}(\mathcal{R})) \right)^{\frac{1}{2}},$$

see Definition 3.1 for the precise definition of the seminorm $\|\cdot\|_{\mathcal{R}}$.

3.2. The seminorm $\|\cdot\|_{\mathcal{R}}$

The main result of this section is Theorem 3.34.

Definition 3.1. We define the vector spaces

$$U_{\mathrm{per},\mathbb{C}} := L^{\infty}_{\mathrm{per}}(\mathcal{G}, \mathbb{C}^{d\times 1}) = \{u\colon \mathcal{G} \to \mathbb{C}^d \,|\, u \text{ is periodic}\}$$

and

$$U_{\mathrm{per}} := \{u\colon \mathcal{G} \to \mathbb{R}^d \,|\, u \text{ is periodic}\} \subset U_{\mathrm{per},\mathbb{C}}.$$

For all $\mathcal{R} \subset \mathcal{G}$ we define the vector spaces

$$U_{\mathrm{trans}}(\mathcal{R}) := \Big\{u\colon \mathcal{R} \to \mathbb{R}^d \,\Big|\, \exists a \in \mathbb{R}^d \,\forall g \in \mathcal{R} : L(g)u(g) = a\Big\},$$

$$U_{\mathrm{rot}}(\mathcal{R}) := \Big\{u\colon \mathcal{R} \to \mathbb{R}^d \,\Big|\, \exists S \in \mathrm{Skew}(d) \,\forall g \in \mathcal{R} : L(g)u(g) = S(g \cdot x_0 - x_0)\Big\}$$

and

$$U_{\mathrm{iso}}(\mathcal{R}) := U_{\mathrm{trans}}(\mathcal{R}) + U_{\mathrm{rot}}(\mathcal{R}).$$

For all finite sets $\mathcal{R} \subset \mathcal{G}$ we define the norm

$$\|\cdot\|\colon \{u\colon \mathcal{R} \to \mathbb{R}^d\} \to [0, \infty)$$

$$u \mapsto \Big(\sum_{g \in \mathcal{R}} \|u(g)\|^2\Big)^{\frac{1}{2}}$$

and the function

$$\|\cdot\|_{\mathcal{R}}\colon U_{\mathrm{per}} \to [0, \infty)$$

$$u \mapsto \Big(\frac{1}{|\mathcal{C}_N|} \sum_{g \in \mathcal{C}_N} \|\pi_{U_{\mathrm{iso}}(\mathcal{R})}(u(g \cdot)|_{\mathcal{R}})\|^2\Big)^{\frac{1}{2}} \quad \text{if } u \text{ is } \mathcal{T}^N\text{-periodic},$$

where $\pi_{U_{\mathrm{iso}}(\mathcal{R})}$ is the orthogonal projection on $\{u\colon \mathcal{R} \to \mathbb{R}^d\}$ with respect to the norm $\|\cdot\|$ with kernel $U_{\mathrm{iso}}(\mathcal{R})$.

Remark 3.2. (i) The map $(\{u\colon \mathcal{R} \to \mathbb{R}^d\}, \|\cdot\|) \to (\mathbb{C}^{d\times|\mathcal{R}|}, \|\cdot\|)$, $u \mapsto (u(g))_{g \in \mathcal{R}}$ is an isomorphism for all finite sets $\mathcal{R} \subset \mathcal{G}$. Thus there is no ambiguity between the above definition and Definition 2.45.

(ii) The definition of $\|\cdot\|_{\mathcal{R}}$ is independent of the choice of $\mathcal{C}_N$ for all $N \in M_0$.

(iii) One could also consider the vector space

$$\Big\{u\colon \mathcal{R} \to \mathbb{R}^d \,\Big|\, \exists S \in \mathrm{Skew}(d) \,\forall g \in \mathcal{R} : L(g)u(g) = S(g \cdot x_0)\Big\},$$

instead of $U_{\mathrm{rot}}(\mathcal{R})$ since its sum with $U_{\mathrm{trans}}(\mathcal{R})$ is also $U_{\mathrm{iso}}(\mathcal{R})$. Due to technical reasons we prefer $U_{\mathrm{rot}}(\mathcal{R})$.

For the definition of a seminorm see Definition C.1.

Lemma 3.3. *For all finite sets $\mathcal{R} \subset \mathcal{G}$ the function $\| \cdot \|_{\mathcal{R}}$ is a seminorm.*

Proof. Let $\mathcal{R} \subset \mathcal{G}$ be finite, $V = \{v \colon \mathcal{G} \to \{w \colon \mathcal{R} \to \mathbb{R}^d\} \,|\, v \text{ is periodic}\}$. We define the maps

$$f_1 \colon U_{\text{per}} \to V$$
$$u \mapsto \left(v \colon \mathcal{G} \to \{w \colon \mathcal{R} \to \mathbb{R}^d\}, \ g \mapsto \pi_{U_{\text{iso}}(\mathcal{R})}(u(g \cdot)|_{\mathcal{R}}) \right)$$

and

$$f_2 \colon V \to [0, \infty)$$
$$v \mapsto \left(\frac{1}{|\mathcal{C}_N|} \sum_{g \in \mathcal{C}_N} \|v(g)\|^2 \right)^{\frac{1}{2}} \quad \text{if } N \in M_0 \text{ and } v \text{ is } \mathcal{T}^N\text{-periodic}.$$

It holds $\| \cdot \|_{\mathcal{R}} = f_2 \circ f_1$. Since f_1 is linear and f_2 is a norm, the map $\| \cdot \|_{\mathcal{R}}$ is a seminorm. $\qquad\square$

Remark 3.4. For all finite sets $\mathcal{R} \subset \mathcal{G}$ the seminorm $\| \cdot \|_{\mathcal{R}}$ satisfies the parallelogram law, i.e. the seminorm is induced by a positive semidefinite symmetric bilinear form.

3.2.1. Equivalence of the seminorms $\| \cdot \|_{\mathcal{R}_1}$ and $\| \cdot \|_{\mathcal{R}_2}$ for appropriate $\mathcal{R}_1, \mathcal{R}_2 \subset \mathcal{G}$

Definition 3.5. We say $\mathcal{R} \subset \mathcal{G}$ has *Property* 1 if $\mathcal{R}$ is finite, $id \in \mathcal{R}$ and

$$\text{aff}(\mathcal{R} \cdot x_0) = \text{aff}(\mathcal{G} \cdot x_0).$$

We say $\mathcal{R} \subset \mathcal{G}$ has *Property* 2 if $\mathcal{R}$ is finite and there exist two sets $\mathcal{R}', \mathcal{R}'' \subset \mathcal{G}$ such that $id \in \mathcal{R}'$, $\mathcal{R}'$ generates $\mathcal{G}$, $\mathcal{R}''$ has Property 1 and $\mathcal{R}'\mathcal{R}'' \subset \mathcal{R}$.

If $\mathcal{R} \subset \mathcal{G}$ has Property 2, then $\mathcal{R}$ has also Property 1.

Lemma 3.6. *Suppose that $\mathcal{R} \subset \mathcal{G}$ has Property 1. Then there exists some $A \in \mathbb{R}^{d_{\text{aff}} \times |\mathcal{R}|}$ of rank d_{aff} such that*

$$(g \cdot x_0 - x_0)_{g \in \mathcal{R}} = \begin{pmatrix} 0_{d - d_{\text{aff}}, |\mathcal{R}|} \\ A \end{pmatrix}.$$

Proof. Suppose that $\mathcal{R} \subset \mathcal{G}$ has Property 1. Since $\mathcal{G} \cdot x_0 \subset \{0_{d-d_{\mathrm{aff}}}\} \times \mathbb{R}^{d_{\mathrm{aff}}}$, there exists some $A \in \mathbb{R}^{d_{\mathrm{aff}} \times |\mathcal{R}|}$ such that

$$(g \cdot x_0 - x_0)_{g \in \mathcal{R}} = \begin{pmatrix} 0 \\ A \end{pmatrix}.$$

It holds

$$\dim(\mathrm{span}(\{g \cdot x_0 - x_0 \mid g \in \mathcal{R}\})) = \dim(\mathrm{aff}(\mathcal{R} \cdot x_0)) = \dim(\mathrm{aff}(\mathcal{G} \cdot x_0)) = d_{\mathrm{aff}}$$

and thus, $\mathrm{rank}(A) = d_{\mathrm{aff}}$. $\square$

Definition 3.7. For all finite sets $\mathcal{R} \subset \mathcal{G}$ we define the function

$$p_{\mathcal{R}} \colon \{u \colon \mathcal{R} \to \mathbb{R}^d\} \to [0, \infty)$$
$$u \mapsto \|\pi_{U_{\mathrm{iso}}(\mathcal{R})}(u)\|,$$

where $\pi_{U_{\mathrm{iso}}(\mathcal{R})}$ is as above.

Lemma 3.8. *For all finite sets $\mathcal{R} \subset \mathcal{G}$ the function $p_{\mathcal{R}}$ is a seminorm and its kernel is $U_{\mathrm{iso}}(\mathcal{R})$.*

Proof. This is clear. $\square$

Definition 3.9. For all finite sets $\mathcal{R}_1, \mathcal{R}_2, \mathcal{R}_3 \subset \mathcal{G}$ such that $\mathcal{R}_1 \subset \mathcal{R}_3 \mathcal{R}_2$, we define the function

$$q_{\mathcal{R}_1, \mathcal{R}_2, \mathcal{R}_3} \colon \{u \colon \mathcal{R}_1 \to \mathbb{R}^d\} \to [0, \infty)$$
$$u \mapsto \inf_{\substack{v \colon \mathcal{R}_3 \mathcal{R}_2 \to \mathbb{R}^d \\ v|_{\mathcal{R}_1} = u}} \left(\sum_{g \in \mathcal{R}_3} p_{\mathcal{R}_2}^2 \big(v(g \cdot)|_{\mathcal{R}_2}\big) \right)^{\frac{1}{2}}.$$

Lemma 3.10. *The infimum in Definition 3.9 is a minimum, i. e. for all finite sets $\mathcal{R}_1, \mathcal{R}_2, \mathcal{R}_3 \subset \mathcal{G}$ such that $\mathcal{R}_1 \subset \mathcal{R}_3 \mathcal{R}_2$ and for all $u \colon \mathcal{R}_1 \to \mathbb{R}^d$ there exists some $v \colon \mathcal{R}_3 \mathcal{R}_2 \to \mathbb{R}^d$ such that $v|_{\mathcal{R}_1} = u$ and*

$$q_{\mathcal{R}_1, \mathcal{R}_2, \mathcal{R}_3}(u) = \left(\sum_{g \in \mathcal{R}_3} p_{\mathcal{R}_2}^2 \big(v(g \cdot)|_{\mathcal{R}_2}\big) \right)^{\frac{1}{2}}.$$

Proof. Let $\mathcal{R}_1, \mathcal{R}_2, \mathcal{R}_3 \subset \mathcal{G}$ be finite such that $\mathcal{R}_1 \subset \mathcal{R}_3 \mathcal{R}_2$. We define the liner map

$$A_g \colon \{v \colon \mathcal{R}_3 \mathcal{R}_2 \to \mathbb{R}^d\} \to \{w \colon \mathcal{R}_2 \to \mathbb{R}^d\}$$
$$v \mapsto \pi_{U_{\mathrm{iso}}(\mathcal{R}_2)}(v(g \cdot)|_{\mathcal{R}_2})$$

for all $g \in \mathcal{R}_2$, where $\pi_{U_{\mathrm{iso}}(\mathcal{R}_2)}$ is as above, and the linear map

$$A \colon \{v \colon \mathcal{R}_3\mathcal{R}_2 \to \mathbb{R}^d\} \to \{w \colon \mathcal{R}_2 \to \mathbb{R}^d\}^{\mathcal{R}_3}$$
$$v \mapsto (A_g v)_{g \in \mathcal{R}_3}.$$

For all $g \in \mathcal{R}_3$ and $v \colon \mathcal{R}_3\mathcal{R}_2 \to \mathbb{R}^d$ we have

$$p_{\mathcal{R}_2}\bigl(v(g \cdot)|_{\mathcal{R}_2}\bigr) = \|A_g v\|$$

and

$$\left(\sum_{g \in \mathcal{R}_3} p_{\mathcal{R}_2}^2\bigl(v(g \cdot)|_{\mathcal{R}_2}\bigr) \right)^{\frac{1}{2}} = \|Av\|.$$

Thus, for all $u \colon \mathcal{R}_1 \to \mathbb{R}^d$ there exists some $v \colon \mathcal{R}_3\mathcal{R}_2 \to \mathbb{R}^d$ such that $v|_{\mathcal{R}_1} = u$ and $q_{\mathcal{R}_1,\mathcal{R}_2,\mathcal{R}_3}(u) = \|Av\|$. $\qquad\square$

Lemma 3.11. *Let $\mathcal{R}_1, \mathcal{R}_2, \mathcal{R}_3 \subset \mathcal{G}$ be finite such that $\mathcal{R}_1 \subset \mathcal{R}_3\mathcal{R}_2$. Then the function $q_{\mathcal{R}_1,\mathcal{R}_2,\mathcal{R}_3}$ is a seminorm.*

Proof. Let $\mathcal{R}_1, \mathcal{R}_2, \mathcal{R}_3 \subset \mathcal{G}$ be finite such that $\mathcal{R}_1 \subset \mathcal{R}_3\mathcal{R}_2$. We have to show that $q_{\mathcal{R}_1,\mathcal{R}_2,\mathcal{R}_3}$ is subadditive. Let $A \colon \{v \colon \mathcal{R}_3\mathcal{R}_2 \to \mathbb{R}^d\} \to \{w \colon \mathcal{R}_2 \to \mathbb{R}^d\}^{\mathcal{R}_3}$ be the map as in the proof of Lemma 3.10. For all $u_1, u_2 \colon \mathcal{R}_1 \to \mathbb{R}^d$ we have

$$
\begin{aligned}
q_{\mathcal{R}_1,\mathcal{R}_2,\mathcal{R}_3}(u_1 + u_2) &= \inf\bigl\{\|Av\| \bigm| v \colon \mathcal{R}_3\mathcal{R}_2 \to \mathbb{R}^d, v|_{\mathcal{R}_1} = u_1 + u_2\bigr\} \\
&= \inf\bigl\{\|Av_1 + Av_2\| \bigm| v_1, v_2 \colon \mathcal{R}_3\mathcal{R}_2 \to \mathbb{R}^d, v_1|_{\mathcal{R}_1} = u_1, v_2|_{\mathcal{R}_2} = u_2\bigr\} \\
&\leq \inf\bigl\{\|Av_1\| + \|Av_2\| \bigm| v_2, v_2 \colon \mathcal{R}_3\mathcal{R}_2 \to \mathbb{R}^d, v_1|_{\mathcal{R}_1} = u_1, v_2|_{\mathcal{R}_1} = u_2\bigr\} \\
&= q_{\mathcal{R}_1,\mathcal{R}_2,\mathcal{R}_3}(u_1) + q_{\mathcal{R}_1,\mathcal{R}_2,\mathcal{R}_3}(u_2). \qquad\square
\end{aligned}
$$

For the definition of the equivalence of two seminorms see Definition C.2

Lemma 3.12. *Suppose that $\mathcal{R}_1 \subset \mathcal{G}$ has Property 1 and $\mathcal{R}_2 \subset \mathcal{G}$ has Property 2. Then there exists a finite set $\mathcal{R}_3 \subset \mathcal{G}$ such that $\mathcal{R}_1 \subset \mathcal{R}_3\mathcal{R}_2$ and the seminorms $p_{\mathcal{R}_1}$ and $q_{\mathcal{R}_1,\mathcal{R}_2,\mathcal{R}_3}$ are equivalent.*

Proof. Suppose that $\mathcal{R}_1 \subset \mathcal{G}$ has Property 1 and $\mathcal{R}_2 \subset \mathcal{G}$ has Property 2. By Lemma 3.8 the map $p_{\mathcal{R}_1}$ is a seminorm with kernel $U_{\mathrm{iso}}(\mathcal{R}_1)$. By Lemma 3.11 for all finite sets $\mathcal{R}_3 \subset \mathcal{G}$ such that $\mathcal{R}_1 \subset \mathcal{R}_3\mathcal{R}_2$ the function $q_{\mathcal{R}_1,\mathcal{R}_2,\mathcal{R}_3}$ is a seminorm. Hence, by Lemma C.4 it suffices to show that there exists a finite set $\mathcal{R}_3 \subset \mathcal{G}$ with $\mathcal{R}_1 \subset \mathcal{R}_3\mathcal{R}_2$ and

$$\ker(q_{\mathcal{R}_1,\mathcal{R}_2,\mathcal{R}_3}) = U_{\mathrm{iso}}(\mathcal{R}_1).$$

First we show that $U_{\mathrm{iso}}(\mathcal{R}_1) \subset \ker(q_{\mathcal{R}_1,\mathcal{R}_2,\mathcal{R}_3})$ for all finite sets $\mathcal{R}_3 \subset \mathcal{G}$ with $\mathcal{R}_1 \subset \mathcal{R}_3\mathcal{R}_2$. Let $\mathcal{R}_3 \subset \mathcal{G}$ with $\mathcal{R}_1 \subset \mathcal{R}_3\mathcal{R}_2$. Let $u \in U_{\mathrm{iso}}(\mathcal{R}_1)$. There exist some $a \in \mathbb{R}^d$ and $S \in \mathrm{Skew}(d)$ such that

$$L(g)u(g) = a + S(g \cdot x_0 - x_0) \qquad \text{for all } g \in \mathcal{R}_1.$$

We define $v \in U_{\mathrm{iso}}(\mathcal{R}_3\mathcal{R}_2)$ by

$$L(g)v(g) = a + S(g \cdot x_0 - x_0) \qquad \text{for all } g \in \mathcal{R}_3\mathcal{R}_2.$$

We have $v|_{\mathcal{R}_1} = u$ and $v(g \cdot)|_{\mathcal{R}_2} \in U_{\mathrm{iso}}(\mathcal{R}_2)$ for all $g \in \mathcal{R}_3$. Using Lemma 3.8 it follows

$$
\begin{aligned}
q^2_{\mathcal{R}_1,\mathcal{R}_2,\mathcal{R}_3}(u) &= \inf_{\substack{w:\ \mathcal{R}_3\mathcal{R}_2 \to \mathbb{R}^d \\ w|_{\mathcal{R}_1} = u}} \sum_{g \in \mathcal{R}_3} p^2_{\mathcal{R}_2}(w(g \cdot)|_{\mathcal{R}_2}) \\
&\leq \sum_{g \in \mathcal{R}_3} p^2_{\mathcal{R}_2}(v(g \cdot)|_{\mathcal{R}_2}) \\
&= 0.
\end{aligned}
$$

Hence, we have $U_{\mathrm{iso}}(\mathcal{R}_1) \subset \ker(q_{\mathcal{R}_1,\mathcal{R}_2,\mathcal{R}_3})$.

Now we show that there exists some $\mathcal{R}_3 \subset \mathcal{G}$ such that $\ker(q_{\mathcal{R}_1,\mathcal{R}_2,\mathcal{R}_3}) \subset U_{\mathrm{iso}}(\mathcal{R}_1)$. By Property 2 of $\mathcal{R}_2$ there exist finite sets $\mathcal{R}_2', \mathcal{R}_2'' \subset \mathcal{G}$ such that $id \in \mathcal{R}_2'$, $\mathcal{R}_2'$ generates $\mathcal{G}$, $\mathcal{R}_2''$ has Property 1 and

$$\mathcal{R}_2'\mathcal{R}_2'' \subset \mathcal{R}_2.$$

Since $\mathcal{R}_2'$ generates $\mathcal{G}$, there exists some $n_0 \in \mathbb{N}$ such that

$$\mathcal{R}_1 \subset \{id\} \cup \bigcup_{k=1}^{n_0} \Big\{ g_1 \ldots g_k \,\Big|\, g_1, \ldots, g_k \in \mathcal{R}_2' \cup (\mathcal{R}_2')^{-1} \Big\}.$$

Let

$$\mathcal{R}_3 = \{id\} \cup \bigcup_{k=1}^{n_0} \Big\{ g_1 \ldots g_k \,\Big|\, g_1, \ldots, g_k \in \mathcal{R}_2' \cup (\mathcal{R}_2')^{-1} \Big\}.$$

Let $u \in \ker(q_{\mathcal{R}_1,\mathcal{R}_2,\mathcal{R}_3})$. By Lemma 3.10 there exists some $v \colon \mathcal{R}_3\mathcal{R}_2 \to \mathbb{R}^d$ such that $v|_{\mathcal{R}_1} = u$ and $p_{\mathcal{R}_2}(v(g \cdot)|_{\mathcal{R}_2}) = 0$ for all $g \in \mathcal{R}_3$. Hence, for all $g \in \mathcal{R}_3$ there exist some $a(g) \in \mathbb{R}^d$ and $S(g) \in \mathrm{Skew}(d)$ such that

$$L(h)v(gh) = a(g) + S(g)(h \cdot x_0 - x_0) \qquad \text{for all } h \in \mathcal{R}_2. \tag{3.4}$$

Since $\mathcal{G} \cdot x_0 \subset \{0_{d-d_{\mathrm{aff}}}\} \times \mathbb{R}^{d_{\mathrm{aff}}}$, we have $h \cdot x_0 - x_0 \in \{0_{d-d_{\mathrm{aff}}}\} \times \mathbb{R}^{d_{\mathrm{aff}}}$ for all $h \in \mathcal{R}_2$. Hence, for all $g \in \mathcal{R}_3$ we may assume

$$S(g) = \begin{pmatrix} 0 & S_1(g) \\ -S_1(g)^{\mathsf{T}} & S_2(g) \end{pmatrix}$$

for some $S_1(g) \in \mathbb{R}^{(d-d_{\mathrm{aff}}) \times d_{\mathrm{aff}}}$ and $S_2(g) \in \mathrm{Skew}(d_{\mathrm{aff}})$. We prove inductively that for $n = 0, 1, \ldots, n_0$ for all $g \in \{id\} \cup \bigcup_{k=1}^{n} \{g_1 \ldots g_k \,|\, g_1, \ldots, g_k \in \mathcal{R}_2' \cup (\mathcal{R}_2')^{-1}\}$ it holds

$$L(g)a(g) = a(id) + S(id)(g \cdot x_0 - x_0) \quad \text{and} \quad S(g) = L(g)^{\mathsf{T}} S(id) L(g). \tag{3.5}$$

For $n = 0$ the induction hypothesis is true.

We assume the induction hypothesis holds for arbitrary but fixed $0 \leq n < n_0$. Let $g \in \{id\} \cup \bigcup_{k=1}^{n} \{g_1 \ldots g_k \,\big|\, g_1, \ldots, g_k \in \mathcal{R}_2' \cup (\mathcal{R}_2')^{-1}\}$ and $r \in \mathcal{R}_2' \cup (\mathcal{R}_2')^{-1}$.

Case 1: $r \in \mathcal{R}_2'$.

Since $g \in \mathcal{R}_3$ and $r\mathcal{R}_2'' \subset \mathcal{R}_2$, by (3.4) we have

$$L(rh)v(grh) = a(g) + S(g)((rh) \cdot x_0 - x_0) \qquad \text{for all } h \in \mathcal{R}_2''. \tag{3.6}$$

Since $gr \in \mathcal{R}_3$ and $\mathcal{R}_2'' \subset \mathcal{R}_2$, by (3.4) we have

$$L(h)v(grh) = a(gr) + S(gr)(h \cdot x_0 - x_0) \qquad \text{for all } h \in \mathcal{R}_2''. \tag{3.7}$$

By (3.6) and (3.7) we have

$$L(r)a(gr) + L(r)S(gr)(h \cdot x_0 - x_0) = a(g) + S(g)((rh) \cdot x_0 - x_0) \tag{3.8}$$

for all $h \in \mathcal{R}_2''$. Since $id \in \mathcal{R}_2''$, by (3.8) we have

$$L(r)a(gr) = a(g) + S(g)(r \cdot x_0 - x_0) \tag{3.9}$$

and with the induction hypothesis follows

$$\begin{aligned} L(gr)a(gr) &= a(id) + S(id)(g \cdot x_0 - x_0) + S(id)L(g)(r \cdot x_0 - x_0) \\ &= a(id) + S(id)((gr) \cdot x_0 - x_0). \end{aligned}$$

By (3.8) and (3.9) we have

$$\begin{aligned} L(r)S(gr)(h \cdot x_0 - x_0) &= S(g)((rh) \cdot x_0 - r \cdot x_0) \\ &= S(g)L(r)(h \cdot x_0 - x_0) \end{aligned} \tag{3.10}$$

for all $h \in \mathcal{R}_2''$. By Lemma 3.6 there exists some $A \in \mathbb{R}^{d_{\text{aff}} \times |\mathcal{R}_2''|}$ of rank d_{aff} such that

$$(h \cdot x_0 - x_0)_{h \in \mathcal{R}_2''} = \begin{pmatrix} 0_{d_{\text{aff}}, |\mathcal{R}_2''|} \\ A \end{pmatrix}.$$

By (3.10) and the induction hypothesis we have

$$(S(gr) - L(gr)^\mathsf{T} S(id) L(gr)) \begin{pmatrix} 0 \\ A \end{pmatrix} = 0. \tag{3.11}$$

By Lemma 2.85 there exist some $B_{gr} \in \mathrm{O}(d - d_{\text{aff}})$ and $C_{gr} \in \mathrm{O}(d_{\text{aff}})$ such that $L(gr) = B_{gr} \oplus C_{gr}$. Equation (3.11) is equivalent to

$$\begin{pmatrix} (S_1(gr) - B_{gr}^\mathsf{T} S_1(id) C_{gr}) A \\ (S_2(gr) - C_{gr}^\mathsf{T} S_2(id) C_{gr}) A \end{pmatrix} = 0.$$

Since the rank of A is equal to the number of its rows, we have $S_1(gr) = B_{gr}^\mathsf{T} S_1(id) C_{gr}$ and $S_2(gr) = C_{gr}^\mathsf{T} S_2(id) C_{gr}$ which is equivalent to $S(gr) = L(gr)^\mathsf{T} S(id) L(gr)$.

Case 2: $r^{-1} \in \mathcal{R}_2'$.

Since $g \in \mathcal{R}_3$ and $\mathcal{R}_2'' \subset \mathcal{R}_2$, by (3.4) we have

$$L(h) v(gh) = a(g) + S(g)(h \cdot x_0 - x_0) \qquad \text{for all } h \in \mathcal{R}_2''. \tag{3.12}$$

Since $gr \in \mathcal{R}_3$ and $r^{-1} \mathcal{R}_2'' \subset \mathcal{R}_2$, by (3.4) we have

$$L(r^{-1} h) v(gh) = a(gr) + S(gr)((r^{-1} h) \cdot x_0 - x_0) \qquad \text{for all } h \in \mathcal{R}_2''. \tag{3.13}$$

By (3.12) and (3.13) we have

$$a(gr) + S(gr)((r^{-1} h) \cdot x_0 - x_0) = L(r)^\mathsf{T} a(g) + L(r)^\mathsf{T} S(g)(h \cdot x_0 - x_0) \tag{3.14}$$

for all $h \in \mathcal{R}_2''$. Since $id \in \mathcal{R}_2''$, by (3.14) we have

$$a(gr) + S(gr)(r^{-1} \cdot x_0 - x_0) = L(r)^\mathsf{T} a(g). \tag{3.15}$$

By (3.14) and (3.15) we have

$$S(gr)((r^{-1} h) \cdot x_0 - x_0) = S(gr)(r^{-1} \cdot x_0 - x_0) + L(r)^\mathsf{T} S(g)(h \cdot x_0 - x_0)$$

for all $h \in \mathcal{R}_2''$. This is equivalent to

$$S(gr) L(r)^\mathsf{T} (h \cdot x_0 - x_0) = L(r)^\mathsf{T} S(g)(h \cdot x_0 - x_0) \tag{3.16}$$

for all $h \in \mathcal{R}_2''$. By Lemma 3.6 there exists some $A \in \mathbb{R}^{d_{\mathrm{aff}} \times |\mathcal{R}_2''|}$ of rank d_{aff} such that

$$(h \cdot x_0 - x_0)_{h \in \mathcal{R}_2''} = \begin{pmatrix} 0_{d_{\mathrm{aff}}, |\mathcal{R}_2''|} \\ A \end{pmatrix}.$$

By (3.16) and the induction hypothesis we have

$$(S(gr) - L(gr)^{\mathsf{T}} S(id) L(gr)) L(r)^{\mathsf{T}} \begin{pmatrix} 0 \\ A \end{pmatrix} = 0. \qquad (3.17)$$

By Lemma 2.85 there exist $B_r, B_{gr} \in O(d - d_{\mathrm{aff}})$ and $C_r, C_{gr} \in O(d_{\mathrm{aff}})$ such that $L(r) = B_r \oplus C_r$ and $L(gr) = B_{gr} \oplus C_{gr}$. Equation (3.17) is equivalent to

$$\begin{pmatrix} (S_1(gr) - B_{gr}^{\mathsf{T}} S_1(id) C_{gr}) C_r^{\mathsf{T}} A \\ (S_2(gr) - C_{gr}^{\mathsf{T}} S_2(id) C_{gr}) C_r^{\mathsf{T}} A \end{pmatrix} = 0.$$

Since C_r is invertible and the rank of A is equal to the number of its rows, we have $S_1(gr) = B_{gr}^{\mathsf{T}} S_1(id) C_{gr}$ and $S_2(gr) = C_{gr}^{\mathsf{T}} S_2(id) C_{gr}$ which is equivalent to $S(gr) = L(gr)^{\mathsf{T}} S(id) L(gr)$. As $S(gr) = L(gr)^{\mathsf{T}} S(id) L(gr)$, we have by (3.15) and the induction hypothesis that

$$\begin{aligned} L(gr)a(gr) &= L(g)a(g) - L(gr)S(gr)(r^{-1} \cdot x_0 - x_0) \\ &= a(id) + S(id)(g \cdot x_0 - x_0) - S(id)L(gr)(r^{-1} \cdot x_0 - x_0) \\ &= a(id) + S(id)((gr) \cdot x_0 - x_0). \end{aligned}$$

Since $\mathcal{R}_1 \subset \mathcal{R}_3$ and $v|_{\mathcal{R}_1} = u$, we have by (3.4) and (3.5) that

$$L(g)u(g) = L(g)v(g) = L(g)a(g) = a(id) + S(id)(g \cdot x_0 - x_0) \quad \text{for all } g \in \mathcal{R}_1$$

and thus, $u \in U_{\mathrm{iso}}(\mathcal{R}_1)$. $\qquad \square$

Theorem 3.13. *Suppose that* $\mathcal{R}_1, \mathcal{R}_2 \subset \mathcal{G}$ *have Property 2. Then the two seminorms* $\| \cdot \|_{\mathcal{R}_1}$ *and* $\| \cdot \|_{\mathcal{R}_2}$ *are equivalent.*

Proof. Suppose that $\mathcal{R}_1, \mathcal{R}_2 \subset \mathcal{G}$ have Property 2. It is sufficient to show that there exists a constant $C > 0$ such that $\| \cdot \|_{\mathcal{R}_1} \leq C \| \cdot \|_{\mathcal{R}_2}$. Property 2 implies Property 1 and thus, by Lemma 3.12 there exists a finite set $\mathcal{R}_3 \subset \mathcal{G}$ such that $\mathcal{R}_1 \subset \mathcal{R}_3 \mathcal{R}_2$ and some $C > 0$ with $p_{\mathcal{R}_1} \leq C q_{\mathcal{R}_1, \mathcal{R}_2, \mathcal{R}_3}$. Let $u \in U_{\mathrm{per}}$. There exists some $N \in M_0$ such that u is

$\mathcal{T}^N$-periodic. We have

$$\|u\|_{\mathcal{R}_1}^2 = \frac{1}{|\mathcal{C}_N|} \sum_{g \in \mathcal{C}_N} p_{\mathcal{R}_1}^2(u(g\,\cdot\,)|_{\mathcal{R}_1})$$

$$\leq \frac{C^2}{|\mathcal{C}_N|} \sum_{g \in \mathcal{C}_N} q_{\mathcal{R}_1,\mathcal{R}_2,\mathcal{R}_3}^2(u(g\,\cdot\,)|_{\mathcal{R}_1})$$

$$= \frac{C^2}{|\mathcal{C}_N|} \sum_{g \in \mathcal{C}_N} \inf_{\substack{v:\, \mathcal{R}_3 \mathcal{R}_2 \to \mathbb{R}^d \\ v|_{\mathcal{R}_1} = u(g\cdot)|_{\mathcal{R}_1}}} \sum_{\tilde{g} \in \mathcal{R}_3} p_{\mathcal{R}_2}^2\bigl(v(\tilde{g}\,\cdot\,)|_{\mathcal{R}_2}\bigr)$$

$$\leq \frac{C^2}{|\mathcal{C}_N|} \sum_{g \in \mathcal{C}_N} \sum_{\tilde{g} \in \mathcal{R}_3} p_{\mathcal{R}_2}^2\bigl(u(g\tilde{g}\,\cdot\,)|_{\mathcal{R}_2}\bigr)$$

$$= \frac{C^2}{|\mathcal{C}_N|} \sum_{\tilde{g} \in \mathcal{R}_3} \sum_{g \in \mathcal{C}_N \tilde{g}} p_{\mathcal{R}_2}^2\bigl(u(g\,\cdot\,)|_{\mathcal{R}_2}\bigr)$$

$$= C^2 |\mathcal{R}_3| \|u\|_{\mathcal{R}_2}^2,$$

where we used that $\mathcal{C}_N \tilde{g}$ is a representation set of $\mathcal{G}/\mathcal{T}^N$ for all $\tilde{g} \in \mathcal{R}_3$ in the last step. Hence, we have $\|\cdot\|_{\mathcal{R}_1} \leq C|\mathcal{R}_3|^{\frac{1}{2}} \|\cdot\|_{\mathcal{R}_2}$. $\qquad\square$

Remark 3.14. In Theorem 3.13 the premise that $\mathcal{R}_1$ and $\mathcal{R}_2$ have Property 2 cannot be weakened to the premise that $\mathcal{R}_1$ and $\mathcal{R}_2$ are generating sets of $\mathcal{G}$ and have Property 1, see Example 3.33.

3.2.2. The seminorms $\|\cdot\|_{\mathcal{R},0}$, $\|\cdot\|_{\mathcal{R},\nabla}$ and $\|\cdot\|_{\mathcal{R},\nabla,0}$

Definition 3.15. For all $\mathcal{R} \subset \mathcal{G}$ we define the vector spaces

$$U_{\mathrm{rot},0}(\mathcal{R}) := \Bigl\{u \colon \mathcal{R} \to \mathbb{R}^d \,\Big|\, \exists S \in \mathrm{Skew}_{0,d_2}(d) \;\forall g \in \mathcal{R} : L(g)u(g) =$$

$$S(g \cdot x_0 - x_0)\Bigr\}$$

$$\subset U_{\mathrm{rot}}(\mathcal{R})$$

and

$$U_{\mathrm{iso},0}(\mathcal{R}) := U_{\mathrm{trans}}(\mathcal{R}) + U_{\mathrm{rot},0}(\mathcal{R}) \subset U_{\mathrm{iso}}(\mathcal{R}),$$

where

$$\mathrm{Skew}_{0,d_2}(d) := \left\{ \begin{pmatrix} S_1 & S_2 \\ -S_2^{\mathsf{T}} & 0 \end{pmatrix} \,\Big|\, S_1 \in \mathrm{Skew}(d_1), S_2 \in \mathbb{R}^{d_1 \times d_2} \right\} \subset \mathrm{Skew}(d).$$

Definition 3.16. For all $u \in U_{\text{per}}$ and finite sets $\mathcal{R} \subset \mathcal{G}$ we define the *discrete derivative*

$$\nabla_{\mathcal{R}} u : \mathcal{G} \to \{v : \mathcal{R} \to \mathbb{R}^d\}$$
$$g \mapsto (\nabla_{\mathcal{R}} u(g) : \mathcal{R} \to \mathbb{R}^d, h \mapsto u(gh) - L(h)^{\mathsf{T}} u(g)).$$

Remark 3.17. Let $\mathcal{R} \subset \mathcal{G}$ be finite, $u \in U_{\text{per}}$ and $v_u : \mathcal{G} \to \mathbb{R}^d, g \mapsto g \cdot (x_0 + u(g))$ which describes the position of the atoms in the physical model, see Section 3.1. Then the relation between the derivative of v_u and the discrete derivative $\nabla_{\mathcal{R}} u$ is given by

$$\big(v_u(gh) - v_u(g)\big)_{h \in \mathcal{R}} = \Big((gh) \cdot x_0 - g \cdot x_0 + L(gh)((\nabla_{\mathcal{R}} u(g))(h))\Big)_{h \in \mathcal{R}}$$

for all $g \in \mathcal{G}$.

If $u \in U_{\text{per}}$ is $\mathcal{T}^N$-periodic for some $N \in M_0$ and $\mathcal{R} \subset \mathcal{G}$ is finite, then also the discrete derivative $\nabla_{\mathcal{R}} u$ is $\mathcal{T}^N$-periodic.

Definition 3.18. For all finite sets $\mathcal{R} \subset \mathcal{G}$ we define the seminorms

$$\| \cdot \|_{\mathcal{R},0} : U_{\text{per}} \to [0, \infty)$$

$$u \mapsto \Big(\frac{1}{|\mathcal{C}_N|} \sum_{g \in \mathcal{C}_N} \|\pi_{U_{\text{iso},0}(\mathcal{R})}(u(g \cdot)|_{\mathcal{R}})\|^2\Big)^{\frac{1}{2}} \quad \text{if } u \text{ is } \mathcal{T}^N\text{-periodic},$$

$$\| \cdot \|_{\mathcal{R},\nabla} : U_{\text{per}} \to [0, \infty)$$

$$u \mapsto \Big(\frac{1}{|\mathcal{C}_N|} \sum_{g \in \mathcal{C}_N} \|\pi_{U_{\text{rot}}(\mathcal{R})}(\nabla_{\mathcal{R}} u(g))\|^2\Big)^{\frac{1}{2}} \quad \text{if } u \text{ is } \mathcal{T}^N\text{-periodic},$$

and
$$\| \cdot \|_{\mathcal{R},\nabla,0} : U_{\text{per}} \to [0, \infty)$$

$$u \mapsto \Big(\frac{1}{|\mathcal{C}_N|} \sum_{g \in \mathcal{C}_N} \|\pi_{U_{\text{rot},0}(\mathcal{R})}(\nabla_{\mathcal{R}} u(g))\|^2\Big)^{\frac{1}{2}} \quad \text{if } u \text{ is } \mathcal{T}^N\text{-periodic},$$

where $\pi_{U_{\text{iso},0}(\mathcal{R})}$, $\pi_{U_{\text{rot}}(\mathcal{R})}$ and $\pi_{U_{\text{rot},0}(\mathcal{R})}$ are the orthogonal projections on $\{u : \mathcal{R} \to \mathbb{R}^d\}$ with respect to the norm $\| \cdot \|$ with kernels $U_{\text{iso},0}(\mathcal{R})$, $U_{\text{rot}}(\mathcal{R})$ and $U_{\text{rot},0}(\mathcal{R})$, respectively.

Remark 3.19.   (i) For all finite sets $\mathcal{R} \subset \mathcal{G}$, the proof that the functions $\| \cdot \|_{\mathcal{R},0}$, $\| \cdot \|_{\mathcal{R},\nabla}$ and $\| \cdot \|_{\mathcal{R},\nabla,0}$ are seminorms is analogous to the proof of Lemma 3.3.

(ii) We have $\| \cdot \|_{\mathcal{R},\nabla} = \| \cdot \|_{\mathcal{R}\setminus\{id\},\nabla}$ and $\| \cdot \|_{\mathcal{R},\nabla,0} = \| \cdot \|_{\mathcal{R}\setminus\{id\},\nabla,0}$ for all finite sets $\mathcal{R} \subset \mathcal{G}$.

(iii) Let $t_i = (I_d, e_i)$ for all $i = 1, \ldots, d$. If $\mathcal{G} = \langle t_1, \ldots, t_d \rangle$ and $\mathcal{R} = \{t_1, \ldots, t_d\}$, then we have $\|\pi_{U_{\text{rot}}(\mathcal{R})}(\nabla_{\mathcal{R}} u(g))\| = \|(\nabla_{\mathcal{R}} u(g) + (\nabla_{\mathcal{R}} u(g))^{\mathsf{T}})/2\|$ for all $u \in U_{\text{per}}$ and $g \in \mathcal{G}$.

Proposition 3.20. *Let $\mathcal{R} \subset \mathcal{G}$ be finite and $id \in \mathcal{R}$. Then the seminorms $\|\cdot\|_{\mathcal{R}}$ and $\|\cdot\|_{\mathcal{R},\nabla}$ are equivalent and the seminorms $\|\cdot\|_{\mathcal{R},0}$ and $\|\cdot\|_{\mathcal{R},\nabla,0}$ are equivalent.*

Proof. Let $\mathcal{R} \subset \mathcal{G}$ be finite, $id \in \mathcal{R}$ and without loss of generality $\mathcal{R} \neq \{id\}$. Let $u \in U_{\text{per}}$. There exists some $N \in M_0$ such that u is $\mathcal{T}^N$-periodic. We have

$$\|u\|_{\mathcal{R},\nabla}^2 = \frac{1}{|\mathcal{C}_N|} \sum_{g \in \mathcal{C}_N} \|\pi_{U_{\text{rot}}(\mathcal{R})}(\nabla_{\mathcal{R}} u(g))\|^2$$

$$\geq \frac{1}{|\mathcal{C}_N|} \sum_{g \in \mathcal{C}_N} \|\pi_{U_{\text{iso}}(\mathcal{R})}(u(g \cdot)|_{\mathcal{R}})\|^2$$

$$= \|u\|_{\mathcal{R}}^2$$

and thus, $\|\cdot\|_{\mathcal{R},\nabla} \geq \|\cdot\|_{\mathcal{R}}$. Let $\mathcal{R}' = \mathcal{R} \setminus \{id\}$. For all $g \in \mathcal{C}_N$ it holds

$$\|\pi_{U_{\text{iso}}(\mathcal{R})}(u(g \cdot)|_{\mathcal{R}})\|^2$$

$$= \inf_{b \in \mathbb{R}^d} \inf_{S \in \text{Skew}(d)} \left\| \left(L(h)u(gh) - b - S(h \cdot x_0 - x_0) \right)_{h \in \mathcal{R}} \right\|^2$$

$$= \inf_{b \in \mathbb{R}^d} \left(\|u(g) - b\|^2 + \inf_{S \in \text{Skew}(d)} \left\| \left(L(h)u(gh) - b \right. \right.$$
$$\left. \left. - S(h \cdot x_0 - x_0) \right)_{h \in \mathcal{R}'} \right\|^2 \right)$$

$$= \inf_{b \in \mathbb{R}^d} \left(\|b\|^2 + \inf_{S \in \text{Skew}(d)} \left\| \left(L(h)u(gh) - u(g) - b \right. \right.$$
$$\left. \left. - S(h \cdot x_0 - x_0) \right)_{h \in \mathcal{R}'} \right\|^2 \right)$$

$$\geq \inf_{b \in \mathbb{R}^d} \left(\|b\|^2 + \frac{1}{|\mathcal{R}'|} \inf_{S \in \text{Skew}(d)} \left\| \left(L(h)u(gh) - u(g) - b \right. \right.$$
$$\left. \left. - S(h \cdot x_0 - x_0) \right)_{h \in \mathcal{R}'} \right\|^2 \right)$$

$$\geq \inf_{b \in \mathbb{R}^d} \left(\|b\|^2 + \frac{1}{|\mathcal{R}'|} \left(\frac{1}{2} \inf_{S \in \text{Skew}(d)} \left\| \left(L(h)u(gh) - u(g) \right. \right. \right.$$
$$\left. \left. \left. - S(h \cdot x_0 - x_0) \right)_{h \in \mathcal{R}'} \right\|^2 - \left\| (b)_{h \in \mathcal{R}'} \right\|^2 \right) \right)$$

$$= \frac{1}{2|\mathcal{R}'|} \left\| \pi_{U_{\mathrm{rot}}(\mathcal{R})} (\nabla_\mathcal{R} u(g)) \right\|^2,$$

where in the second to last step we used that $\|v + w\|^2 \geq \|v\|^2/2 - \|w\|^2$ for all $v, w \colon \mathcal{R}' \to \mathbb{R}^d$. Thus, we have

$$
\begin{aligned}
\|u\|_\mathcal{R}^2 &= \frac{1}{|\mathcal{C}_N|} \sum_{g \in \mathcal{C}_N} \left\| \pi_{U_{\mathrm{iso}}(\mathcal{R})} (u(g \cdot)|_\mathcal{R}) \right\|^2 \\
&\geq \frac{1}{2|\mathcal{R}'||\mathcal{C}_N|} \sum_{g \in \mathcal{C}_N} \left\| \pi_{U_{\mathrm{rot}}(\mathcal{R})} (\nabla_\mathcal{R} u(g)) \right\|^2 \\
&= \frac{1}{2|\mathcal{R}'|} \|u\|_{\mathcal{R}, \nabla}^2.
\end{aligned}
$$

Hence, we have $\| \cdot \|_{\mathcal{R}, \nabla} \leq \sqrt{2|\mathcal{R}'|} \| \cdot \|_\mathcal{R}$.
The proof of the equivalence of the seminorms $\| \cdot \|_{\mathcal{R}, 0}$ and $\| \cdot \|_{\mathcal{R}, \nabla, 0}$ is analogous. $\qquad \square$

3.2.3. Equivalence of the seminorms $\| \cdot \|_\mathcal{R}$ and $\| \cdot \|_{\mathcal{R}, 0}$

The following lemma is well-known.

Lemma 3.21. *There exists a constant $c > 0$ such that for every $n \in \mathbb{N}$ it holds*

$$\left\| x \otimes y^\mathsf{T} + A \right\| \geq c \left(\left\| x \otimes y^\mathsf{T} \right\| + \|A\| \right) \qquad \text{for all } x, y \in \mathbb{C}^n,\ A \in \mathrm{Skew}(n, \mathbb{C}).$$

Proof. Let $x, y \in \mathbb{C}^n$ and $A \in \mathrm{Skew}(n, \mathbb{C})$. Since $\mathbb{C}^{n \times n} = \mathrm{Sym}(n, \mathbb{C}) \oplus \mathrm{Skew}(n, \mathbb{C})$ we have

$$
\begin{aligned}
\left\| x \otimes y^\mathsf{T} + A \right\|^2 &\geq \left\| \frac{1}{2} (x \otimes y^\mathsf{T} + y \otimes x^\mathsf{T}) \right\|^2 \\
&= \frac{1}{2} \left\| x \otimes y^\mathsf{T} \right\|^2 + \frac{1}{2} \left(\sum_{i=1}^n x_i y_i \right)^2 \\
&\geq \frac{1}{2} \left\| x \otimes y^\mathsf{T} \right\|^2.
\end{aligned}
$$

If $\|A\| \leq 2 \|x \otimes y^\mathsf{T}\|$, then

$$\left\| x \otimes y^\mathsf{T} + A \right\| \geq \frac{1}{\sqrt{2}} \left\| x \otimes y^\mathsf{T} \right\| \geq \frac{1}{3\sqrt{2}} \left(\left\| x \otimes y^\mathsf{T} \right\| + \|A\| \right).$$

If $\|A\| \geq 2 \|x \otimes y^\mathsf{T}\|$, then

$$\left\| x \otimes y^\mathsf{T} + A \right\| \geq \|A\| - \left\| x \otimes y^\mathsf{T} \right\| \geq \frac{1}{3} \left(\left\| x \otimes y^\mathsf{T} \right\| + \|A\| \right). \qquad \square$$

For the proof of Theorem 3.24 we need the following lemma.

Lemma 3.22. *Let $n \in \mathbb{N}$, $q \in \mathbb{N}_0$ and $\beta_1, \ldots, \beta_q \in \mathbb{R}$. Then there exists an integer $N \in \mathbb{N}$ such that*

$$\max_{m \in \{1,\ldots,N\}} \left\| a \otimes (\sin(m\alpha_1), \ldots, \sin(m\alpha_n)) + \sum_{k=1}^{q} \sin(m\beta_k) B_k + mS \right\| \geq \|S\|$$

for all $a \in \mathbb{C}^n$, $\alpha_1, \ldots, \alpha_n \in \mathbb{R}$, $B_1, \ldots, B_q \in \mathbb{C}^{n \times n}$ and $S \in \mathrm{Skew}(n, \mathbb{C})$.

Remark 3.23. If $q = 0$, then the term $\sum_{k=1}^{q} \sin(m\beta_k) B_k$ is the empty sum.

Proof. It suffices to prove that there exists a constant $c > 0$ such that for all $n \in \mathbb{N}$, $q \in \mathbb{N}_0$ and $\beta_1, \ldots, \beta_q \in \mathbb{R}$ there exists an integer $N \in \mathbb{N}$ such that

$$\max_{m \in \{1,\ldots,N\}} \left\| a \otimes (\sin(m\alpha_1), \ldots, \sin(m\alpha_n)) + \sum_{k=1}^{q} \sin(m\beta_k) B_k + mS \right\| \geq c\|S\|$$

for all $a \in \mathbb{C}^n$, $\alpha_1, \ldots, \alpha_n \in \mathbb{R}$, $B_1, \ldots, B_q \in \mathbb{C}^{n \times n}$ and $S \in \mathrm{Skew}(n, \mathbb{C})$ due to the fact that for $N = \lceil \frac{1}{c} \rceil \tilde{N}$ we have

$$\max_{m \in \{1,\ldots,N\}} \left\| a \otimes (\sin(m\alpha_1), \ldots, \sin(m\alpha_n)) + \sum_{k=1}^{q} \sin(m\beta_k) B_k + mS \right\|$$

$$\geq \max_{m \in \left\{1,\ldots,\tilde{N}\right\}} \left\| a \otimes (\sin(m(\lceil \tfrac{1}{c} \rceil \alpha_1)), \ldots, \sin(m(\lceil \tfrac{1}{c} \rceil \alpha_n))) \right.$$

$$\left. + \sum_{k=1}^{q} \sin(m(\lceil \tfrac{1}{c} \rceil \beta_k)) B_k + m(\lceil \tfrac{1}{c} \rceil S) \right\|.$$

Since

$$\|M\| \geq \frac{1}{n^2} \sum_{\substack{i,j \in \{1,\ldots,n\} \\ i<j}} \left\| \left(\begin{smallmatrix} m_{ii} & m_{ij} \\ m_{ji} & m_{jj} \end{smallmatrix} \right) \right\|$$

for all $M = (m_{ij}) \in \mathbb{C}^{n \times n}$, it suffices to prove the assertion for $n = 2$. Let $q \in \mathbb{N}_0$ and $\beta_1, \ldots, \beta_q \in \mathbb{R}$. Without loss of generality we assume $\beta_1, \ldots, \beta_q \in \mathbb{R} \setminus (\pi\mathbb{Q})$: Let $n_0 \in \mathbb{N}$ be such that $n_0\beta_k \in \pi\mathbb{Z}$ for all

$k \in \{1, \ldots, q\}$ with $\beta_k \in \pi\mathbb{Q}$. Then we have

$$
\max_{m \in \{1, \ldots, n_0 N\}} \left\| a \otimes (\sin(m\alpha_1), \sin(m\alpha_2)) + \sum_{k=1}^{q} \sin(m\beta_k) B_k + mS \right\|
$$

$$
\geq \max_{m \in \{1, \ldots, N\}} \left\| a \otimes (\sin(m(n_0\alpha_1)), \sin(m(n_0\alpha_2))) \right.
$$

$$
\left. + \sum_{\substack{k=1 \\ \beta_k \notin \pi\mathbb{Q}}}^{q} \sin(m(n_0\beta_k)) B_k + m(n_0 S) \right\|
$$

for all $N \in \mathbb{N}$, $a \in \mathbb{C}^2$, $\alpha_1, \alpha_2 \in \mathbb{R}$, $B_1, \ldots, B_q \in \mathbb{C}^{2 \times 2}$ and $S \in \mathrm{Skew}(2, \mathbb{C})$. For all $a > 0$ we define the function

$$
| \cdot |_a \colon \mathbb{R} \to [0, \infty)
$$
$$
x \mapsto \mathrm{dist}(x, a\mathbb{Z}).
$$

Moreover, without loss of generality we may assume $|\beta_k - \beta_l|_{2\pi} > 0$ for all $k \neq l$ and since

$$
\sin(m\beta) = -\sin(m(2\pi - \beta))
$$

also $|\beta_k + \beta_l|_{2\pi} \neq 0$ for all $k \neq l$. For the definition of a suitable integer $N \in \mathbb{N}$ and the following proof we define some positive constants. By Lemma 3.21 there exists a constant $c_L > 0$ such that

$$
\| x \otimes y^{\mathsf{T}} + S \| \geq c_L \|x\| (|y_1| + |y_2|) + c_L \|S\|
$$

for all $x, y \in \mathbb{C}^2$ and $S \in \mathrm{Skew}(2, \mathbb{C})$. In particular, this inequality implies the assertion for $q = 0$. Hence we may assume $q \neq 0$, i.e. $q \in \mathbb{N}$. Let

$$
\delta_1 = \min_{\substack{\gamma_1, \gamma_2 \in \{\pm\beta_1, \ldots, \pm\beta_q\} \\ \gamma_1 \neq \gamma_2}} |\gamma_1 - \gamma_2|_{2\pi}, \quad \mu_1 = \frac{1}{2q} \left(\frac{\delta_1}{2\pi} \right)^{2q-1},
$$

$$
C_1 = \frac{4(2q+1)}{\mu_1}, \quad C_2 = \frac{6q}{\mu_1} \quad \text{and} \quad C_3 = \max\left\{ \frac{4q+2}{\mu_1}, \frac{32\pi C_2}{\delta_1} \right\}.
$$

By Kronecker's approximation Theorem D.5, for all $k \in \{1, \ldots, q\}$ there exists an integer q_k such that $2C_3 + 2 < q_k$ and

$$
\left| q_k \cdot \frac{\beta_k}{\pi} + \frac{1}{2} \right|_1 \leq \frac{1}{3\pi C_3}.
$$

Let
$$N_1 = \max\left\{ \left\lceil \frac{2C_1}{c_L} \right\rceil, 2q, 1 + \left\lceil \frac{16\pi C_2}{\delta_1} \right\rceil, q_1, \ldots, q_q \right\} \in \mathbb{N}.$$

For all $\alpha \in \mathbb{R}$ we define $(\alpha)_{2\pi} \in \mathbb{R}$ by $\{(\alpha)_{2\pi}\} = [-\pi, \pi) \cap (\alpha + 2\pi\mathbb{Z})$. We have $|(\alpha)_{2\pi}| = |\alpha|_{2\pi}$. By Taylor's Theorem we have for all $\alpha, \beta \in \mathbb{R}$ and $n \in \mathbb{N}$

$$\sin(n\alpha) = \sin(n(\beta + (\alpha - \beta)_{2\pi})) = \sin(n\beta) + n(\alpha - \beta)_{2\pi}\cos(n\beta) + R(n, \alpha, \beta)$$

where $R(n, \alpha, \beta)$ is the remainder term. Let $\delta_2 > 0$ be so small such that

$$|R(n, \alpha, \beta)| \leq \tfrac{1}{2}n|\alpha - \beta|_{2\pi}|\cos(n\beta)| \tag{3.18}$$

for all $n \in \{1, \ldots, N_1\}$, $\alpha \in \mathbb{R}$ with $|\alpha - \beta|_{2\pi} < \delta_2$ and $\beta \in \{0, \pi, \beta_1, \ldots, \beta_q\}$. Let

$$\delta_3 = \min\{\delta_1, \delta_2\}, \quad \mu_2 = \frac{1}{2q+2}\left(\frac{\delta_3}{2\pi}\right)^{2q+1} \quad \text{and} \quad C_4 = \frac{2q+3}{\mu_2}.$$

Let
$$N = \max\{N_1, 1 + \lceil C_4 \rceil\} \in \mathbb{N}.$$

Now, let $a = (a_1, a_2)^\mathsf{T} \in \mathbb{C}^2$, $\alpha_1, \alpha_2 \in \mathbb{R}$, $B_k = \begin{pmatrix} b_{11}^{(k)} & b_{12}^{(k)} \\ b_{21}^{(k)} & b_{22}^{(k)} \end{pmatrix} \in \mathbb{C}^{2\times 2}$ for all $k \in \{1, \ldots, q\}$ and $S = \begin{pmatrix} 0 & -s \\ s & 0 \end{pmatrix} \in \mathrm{Skew}(2, \mathbb{C})$. We denote

$$\text{LHS} = \max_{m \in \{1, \ldots, N\}}\left\| a \otimes (\sin(m\alpha_1), \sin(m\alpha_2)) + \sum_{k=1}^{q} \sin(m\beta_k)B_k + mS \right\|.$$

Case 1: $\forall i \in \{1, 2\} : ((|\alpha_i|_{2\pi} < \delta_2) \vee (|\alpha_i - \pi|_{2\pi} < \delta_2))$.

Case 1.1: $\sum_{k=1}^{q} \|B_k\| \geq C_1(\|a\|(|\alpha_1|_\pi + |\alpha_2|_\pi) + \|S\|)$.

Let $i, j \in \{1, 2\}$ with $\sum_{k=1}^{q} |b_{ij}^{(k)}| \geq \frac{1}{4}\sum_{k=1}^{q} \|B_k\|$. By the definition of δ_1 we have

$$\min_{\substack{\gamma_1, \gamma_2 \in \{\pm\beta_1, \ldots, \pm\beta_q\} \\ \gamma_1 \neq \gamma_2}} |e^{i\gamma_1} - e^{i\gamma_2}| \geq \min_{\substack{\gamma_1, \gamma_2 \in \{\pm\beta_1, \ldots, \pm\beta_q\} \\ \gamma_1 \neq \gamma_2}} \frac{|\gamma_1 - \gamma_2|_{2\pi}}{\pi} \geq \frac{\delta_1}{\pi}.$$

By Turán's third Theorem D.6 there exists some $\nu \in \{1, \ldots, 2q\}$ such that

$$\left\| \sum_{k=1}^{q} \sin(\nu\beta_k)B_k \right\| \geq \left| \sum_{k=1}^{q} b_{ij}^{(k)}\sin(\nu\beta_k) \right|$$

$$= \left| \sum_{k=1}^{q}\left(\frac{ib_{ij}^{(k)}}{2}e^{-i\nu\beta_k} + \frac{-ib_{ij}^{(k)}}{2}e^{i\nu\beta_k} \right) \right|$$

$$\geq \mu_1 \sum_{k=1}^{q} |b_{ij}^{(k)}|$$

$$\geq \frac{\mu_1}{4} \sum_{k=1}^{q} \|B_k\|.$$

We have

$$\text{LHS} \geq \left\| \sum_{k=1}^{q} \sin(\nu\beta_k)B_k \right\| - \|a \otimes (\sin(\nu\alpha_1), \sin(\nu\alpha_2))\| - \|\nu S\|$$

$$\geq \frac{\mu_1}{4} \sum_{k=1}^{q} \|B_k\| - 2q\|a\|(|\alpha_1|_\pi + |\alpha_2|_\pi) - 2q\|S\|$$

$$\geq \|S\|.$$

Case 1.2: $\sum_{k=1}^{q} \|B_k\| \leq C_1(\|a\|(|\alpha_1|_\pi + |\alpha_2|_\pi) + \|S\|)$.
We have

$$\text{LHS} \geq \|a \otimes (\sin(N_1\alpha_1), \sin(N_1\alpha_2)) + N_1 S\| - \left\| \sum_{k=1}^{q} \sin(N_1\beta_k)B_k \right\|$$

$$\geq c_L\|a\|(|\sin(N_1\alpha_1)| + |\sin(N_1\alpha_2)|) + c_L\|N_1 S\| - \sum_{k=1}^{q} \|B_k\|$$

$$\overset{(3.18)}{\geq} \frac{c_L N_1}{2}\|a\|(|\alpha_1|_\pi + |\alpha_2|_\pi) + c_L N_1\|S\| - \sum_{k=1}^{q} \|B_k\|$$

$$\geq \frac{c_L N_1}{2}(\|a\|(|\alpha_1|_\pi + |\alpha_2|_\pi) + \|S\|) + \frac{c_L}{2}\|S\| - \sum_{k=1}^{q} \|B_k\|$$

$$\geq \frac{c_L}{2}\|S\|.$$

Case 2: $\exists i \in \{1,2\}, \exists k \in \{1,\ldots,q\} : ((|\alpha_i - \beta_k|_{2\pi} < \delta_2) \vee$
$(|\alpha_i + \beta_k|_{2\pi} < \delta_2))$.
Without loss of generality let $i = 1$ and $k = 1$. Without loss of generality we may assume $|\alpha_1 - \beta_1|_{2\pi} < \delta_2$ since

$$a \otimes (\sin(m\alpha_1), \sin(m\alpha_2)) = (-a) \otimes (\sin(m(-\alpha_1)), \sin(m(-\alpha_2)))$$

for all $m \in \mathbb{N}$. Let δ_k be equal to 1 if $k = 0$ and equal to 0 otherwise.

Case 2.1: $\sum_{k=1}^{q} |a_2\delta_{k-1} + b_{21}^{(k)}| \geq C_2|a_2||\alpha_1 - \beta_1|_{2\pi}$ *and*
$\max\{|a_2||\alpha_1 - \beta_1|_{2\pi}, \sum_{k=1}^{q} |a_2\delta_{k-1} + b_{21}^{(k)}|\} \geq C_3|s|$.

Since $C_2 \geq 1$ the condition is equivalent to

$$\sum_{k=1}^{q}|a_2\delta_{k-1}+b_{21}^{(k)}| \geq C_2|a_2||\alpha_1-\beta_1|_{2\pi} \text{ and } \sum_{k=1}^{q}|a_2\delta_{k-1}+b_{21}^{(k)}| \geq C_3|s|.$$

By Turán's third theorem (analogously to Case 1.1) there exists an integer $\nu \in \{1,\ldots,2q\}$ such that

$$\left|\sum_{k=1}^{q}(a_2\delta_{k-1}+b_{21}^{(k)})\sin(\nu\beta_k)\right|$$

$$=\left|\sum_{k=1}^{q}\left(\frac{\mathrm{i}(a_2\delta_{k-1}+b_{21}^{(k)})}{2}\mathrm{e}^{-\mathrm{i}\nu\beta_k} + \frac{-\mathrm{i}(a_2\delta_{k-1}+b_{21}^{(k)})}{2}\mathrm{e}^{\mathrm{i}\nu\beta_k}\right)\right|$$

$$\geq \mu_1\sum_{k=1}^{q}|a_2\delta_{k-1}+b_{21}^{(k)}|.$$

We have

$$\text{LHS} \overset{(3.18)}{\geq} \left|\sum_{k=1}^{q}(a_2\delta_{k-1}+b_{21}^{(k)})\sin(\nu\beta_k)\right|$$

$$-\frac{3}{2}|a_2||\nu||\alpha_1-\beta_1|_{2\pi}|\cos(\nu\beta_1)|-\nu|s|$$

$$\geq \left(\frac{\mu_1}{2}+\frac{\mu_1}{2}\right)\sum_{k=1}^{q}|a_2\delta_{k-1}+b_{21}^{(k)}| - 3q|a_2||\alpha_1-\beta_1|_{2\pi} - 2q|s|$$

$$\geq |s|$$

$$=\frac{1}{\sqrt{2}}\|S\|.$$

Case 2.2: $\sum_{k=1}^{q}|a_2\delta_{k-1}+b_{21}^{(k)}| \leq C_2|a_2||\alpha_1-\beta_1|_{2\pi}$ *and*
$\max\{|a_2||\alpha_1-\beta_1|_{2\pi}, \sum_{k=1}^{q}|a_2\delta_{k-1}+b_{21}^{(k)}|\} \geq C_3|s|.$
By Turán's third theorem there exists an integer $\nu \in \{N_1-1,N_1\}$ such that

$$|\cos(\nu\beta_1)| = \left|\tfrac{1}{2}\mathrm{e}^{\mathrm{i}\nu\beta_1} + \tfrac{1}{2}\mathrm{e}^{-\mathrm{i}\nu\beta_1}\right| \geq \frac{\delta_1}{4\pi}.$$

We have

$$\text{LHS} \overset{(3.18)}{\geq} \frac{1}{2}|a_2||\cos(\nu\beta_1)||\nu|\alpha_1$$

$$-\beta_1|_{2\pi} - \left|\sum_{k=1}^{q}(a_2\delta_{k-1}+b_{21}^{(k)})\sin(\nu\beta_k)\right| - \nu|s|$$

$$\geq \left(\frac{\delta_1(N_1 - 1)}{16\pi} + \frac{\delta_1 \nu}{16\pi} \right) |a_2||\alpha_1 - \beta_1|_{2\pi}$$

$$- \sum_{k=1}^{q} |a_2 \delta_{k-1} + b_{21}^{(k)}| - \nu|s|$$

$$\geq \nu|s|$$

$$\geq \frac{1}{\sqrt{2}} \|S\|.$$

Case 2.3: $\max\{|a_2||\alpha_1 - \beta_1|_{2\pi}, \sum_{k=1}^{q} |a_2 \delta_{k-1} + b_{21}^{(k)}|\} \leq C_3 |s|.$
By Definition of q_1 we have

$$|\cos(q_1 \beta_1)| = |\sin(q_1 \beta_1 + \tfrac{\pi}{2})| \leq |q_1 \beta_1 + \tfrac{\pi}{2}|_{\pi} = \pi |\tfrac{q_1 \beta_1}{\pi} + \tfrac{1}{2}|_1 \leq \tfrac{1}{3C_3}.$$

So we have

$$\text{LHS} \overset{(3.18)}{\geq} q_1|s| - \tfrac{3}{2}|a_2||\cos(q_1 \beta_1)|q_1|\alpha_1 - \beta_1|_{2\pi}$$

$$- \left| \sum_{k=1}^{q} \left(a_2 \delta_{k-1} + b_{21}^{(k)} \right) \sin(q_1 \beta_k) \right|$$

$$\geq \left(1 + \frac{q_1}{2} + C_3 \right)|s| - \frac{q_1}{2C_3}|a_2||\alpha_1 - \beta_1|_{2\pi}$$

$$- \sum_{k=1}^{q} |a_2 \delta_{k-1} + b_{21}^{(k)}|$$

$$\geq \frac{1}{\sqrt{2}} \|S\|.$$

Case 3: $\exists i \in \{1, 2\} : (|\alpha_i - \beta|_{2\pi} \geq \delta_2 \quad \forall \beta \in \{0, \pi, \pm\beta_1, \ldots, \pm\beta_q\}).$
Without loss of generality let $i = 1$.

Case 3.1: $|a_2| + \sum_{k=1}^{q} |b_{21}^{(k)}| \geq C_4 |s|.$
By Definition of δ_3 we have

$$\min_{\substack{\gamma_1, \gamma_2 \in \{\pm\alpha_1, \pm\beta_1, \ldots, \pm\beta_q\} \\ \gamma_1 \neq \gamma_2}} |e^{i\gamma_1} - e^{i\gamma_2}|$$

$$\geq \min_{\substack{\gamma_1, \gamma_2 \in \{\pm\alpha_1, \pm\beta_1, \ldots, \pm\beta_q\} \\ \gamma_1 \neq \gamma_2}} \frac{|\gamma_1 - \gamma_2|_{2\pi}}{\pi}$$

$$\geq \frac{\min\{\delta_1, \delta_2\}}{\pi}$$

$$= \frac{\delta_3}{\pi}.$$

By Turán's third theorem there exists an integer $\nu \in \{1, \ldots, 2q+2\}$ such that

$$\left| a_2 \sin(\nu\alpha_1) + \sum_{k=1}^{q} b_{21}^{(k)} \sin(\nu\beta_k) \right|$$

$$= \left| \frac{\mathrm{i}a_2}{2}\mathrm{e}^{-\mathrm{i}\nu\alpha_1} + \frac{-\mathrm{i}a_2}{2}\mathrm{e}^{\mathrm{i}\nu\alpha_1} + \sum_{k=1}^{q} \left(\frac{\mathrm{i}b_{21}^{(k)}}{2}\mathrm{e}^{-\mathrm{i}\nu\beta_k} + \frac{-\mathrm{i}b_{21}^{(k)}}{2}\mathrm{e}^{\mathrm{i}\nu\beta_k} \right) \right|$$

$$\geq \mu_2 \left(|a_2| + \sum_{k=1}^{q} |b_{21}^{(k)}| \right).$$

We have

$$\text{LHS} \geq \left| a_2 \sin(\nu\alpha_1) + \sum_{k=1}^{q} b_{21}^{(k)} \sin(\nu\beta_k) \right| - \nu|s|$$

$$\geq \mu_2 \left(|a_2| + \sum_{k=1}^{q} |b_{21}^{(k)}| \right) - (2q+2)|s|$$

$$\geq |s|$$

$$= \frac{1}{\sqrt{2}}\|S\|.$$

Case 3.2: $|a_2| + \sum_{k=1}^{q} |b_{21}^{(k)}| \leq C_4|s|.$

We have

$$\text{LHS} \geq N|s| - |a_2 \sin(N\alpha_1)| - \left| \sum_{k=1}^{q} b_{21}^{(k)} \sin(N\beta_k) \right|$$

$$\geq N|s| - \left(|a_2| + \sum_{k=1}^{q} |b_{21}^{(k)}| \right)$$

$$\geq |s|$$

$$= \frac{1}{\sqrt{2}}\|S\|.$$

Since case 2 and case 3 include the case

$$\exists i \in \{1,2\} : ((|\alpha_i|_{2\pi} \geq \delta_2) \wedge (|\alpha_i - \pi|_{2\pi} \geq \delta_2)),$$

the assertion is proven. $\qquad\square$

Theorem 3.24 (A discrete Korn's inequality). *Suppose that* $\mathcal{R} \subset \mathcal{G}$ *has Property 2. Then the two seminorms* $\|\cdot\|_{\mathcal{R}}$ *and* $\|\cdot\|_{\mathcal{R},0}$ *are equivalent.*

Proof. Suppose that $\mathcal{R} \subset \mathcal{G}$ has Property 2.

First we show the trivial inequality $\| \cdot \|_{\mathcal{R}} \leq \| \cdot \|_{\mathcal{R},0}$:

Let $u \in U_{\text{per}}$. Let $N \in M_0$ be such that u is T^N-periodic. Since $U_{\text{iso},0}(\mathcal{R}) \subset U_{\text{iso}}(\mathcal{R})$, we have

$$\|u\|_{\mathcal{R}}^2 = \frac{1}{|\mathcal{C}_N|} \sum_{g \in \mathcal{C}_N} \|\pi_{U_{\text{iso}}(\mathcal{R})}(u(g \cdot)|_{\mathcal{R}})\|^2$$

$$\leq \frac{1}{|\mathcal{C}_N|} \sum_{g \in \mathcal{C}_N} \|\pi_{U_{\text{iso},0}(\mathcal{R})}(u(g \cdot)|_{\mathcal{R}})\|^2$$

$$= \|u\|_{\mathcal{R},0}^2.$$

Now we show with the aid of the Plancherel formula that there exists a constant $c > 0$ such that $\| \cdot \|_{\mathcal{R}} \geq c\| \cdot \|_{\mathcal{R},0}$:

By Theorem 2.17 there exists some $m \in \mathbb{N}$ such that $M_0 = m\mathbb{N}$. By Lemma 2.12 the group T^m is isomorphic to $\mathbb{Z}^{d_2}$ and thus, there exist $t_1, \ldots, t_{d_2} \in T^m$ such that $\{t_1, \ldots, t_{d_2}\}$ generates T^m. Since $L(T^m)$ is a subgroup of $\oplus(\mathrm{O}(d - d_{\text{aff}}) \times \mathrm{O}(d_{\text{aff}} - d_2) \times \{I_{d_2}\})$ and the elements $t_1, \ldots, t_{d_2}$ commute, by Theorem D.4 we may without loss of generality (by a coordinate transformation) assume that there exist matrices $A_1, \ldots, A_{d_2}$, an integer $q \in \{0, \ldots, \lfloor(d_{\text{aff}} - d_2)/2\rfloor\}$, vectors $v_1, \ldots, v_{d_2} \in \{\pm 1\}^{d_{\text{aff}} - d_2 - 2q}$ and angles $\theta_{1,1}, \ldots, \theta_{d_2,q} \in [0, 2\pi)$ such that

$$L(t_i) = A_i \oplus \mathrm{diag}(v_i) \oplus R(\theta_{i,1}) \oplus \cdots \oplus R(\theta_{i,q}) \oplus I_{d_2} \quad \text{for all } i \in \{1, \ldots, d_2\}.$$

By Lemma 3.22 there exists an integer $N_0 \in \mathbb{N}$ such that

$$\max_{n \in \{1, \ldots, N_0\}} \left\| a \otimes (\sin(n\alpha_1), \ldots, \sin(n\alpha_{d_2})) - \sum_{i=1}^{d_2} \sum_{j=1}^{q} \sin(n\theta_{i,j}) B_{i,j} - nS \right\|$$

$$\geq \|S\| \tag{3.19}$$

for all $a \in \mathbb{C}^{d_2}$, $\alpha_1, \ldots, \alpha_{d_2} \in [0, 2\pi)$, $B_{1,1}, \ldots, B_{d_2,q} \in \mathbb{C}^{d_2 \times d_2}$, and $S \in \mathrm{Skew}(d_2, \mathbb{C})$. Let $\mathcal{R}_0 = \{t_i^n \mid i \in \{1, \ldots, d_2\}, n \in \{\pm 1, \ldots, \pm N_0\}\} \subset T^m$. Since $\| \cdot \|_{\mathcal{R} \cup \mathcal{R}_0, 0} \geq \| \cdot \|_{\mathcal{R},0}$ and by Theorem 3.13, we may without loss of generality assume that $\mathcal{R}_0 \subset \mathcal{R}$. For all finite sets $\mathcal{R}' \subset \mathcal{G}$ we define the map

$$g_{\mathcal{R}'} \colon \mathrm{Skew}(d, \mathbb{C}) \to \mathbb{C}^{d \times |\mathcal{R}'|}$$
$$S \mapsto (L(h)^{\mathsf{T}} S(h \cdot x_0 - x_0))_{h \in \mathcal{R}'}.$$

Now we show that there exists a constant $c_0 > 0$ such that

$$\left\| (\chi(h)v - L(h)^{\mathsf{T}}v)_{h \in \mathcal{R}_0} - g_{\mathcal{R}_0}(S) \right\| \geq c_0 \|S_3\| \tag{3.20}$$

for all $\chi \in \widehat{\mathcal{T}^m}$, $v \in \mathbb{C}^d$ and $S = \begin{pmatrix} S_1 & -S_2^{\mathsf{T}} \\ S_2 & S_3 \end{pmatrix} \in \text{Skew}(d_1 + d_2, \mathbb{C})$.

Let $\chi \in \widehat{\mathcal{T}^m}$, $v = \begin{pmatrix} v_1 \\ v_2 \end{pmatrix} \in \mathbb{C}^{d_1+d_2}$ and $S = \begin{pmatrix} S_1 & -S_2^{\mathsf{T}} \\ S_2 & S_3 \end{pmatrix} \in \text{Skew}(d_1 + d_2, \mathbb{C})$. We have

$$
\begin{aligned}
\text{LHS} &:= \left\| \left(\chi(h)v - L(h)^{\mathsf{T}}v \right)_{h \in \mathcal{R}_0} - g_{\mathcal{R}_0}(S) \right\| \\
&\geq \left\| \left(\chi(h)v_2 - v_2 - (S_2, S_3)(h \cdot x_0 - x_0) \right)_{h \in \mathcal{R}_0} \right\| \\
&\geq \frac{1}{\sqrt{2}} \Bigg(\left\| \left(\chi(t_i^n)v_2 - v_2 - (S_2, S_3)(t_i^n \cdot x_0 - x_0) \right)_{i \in \{1,\ldots,d_2\}} \right\| \\
&\quad + \left\| \left(\chi(t_i^{-n})v_2 - v_2 - (S_2, S_3)(t_i^{-n} \cdot x_0 - x_0) \right)_{i \in \{1,\ldots,d_2\}} \right\| \Bigg) \\
&\geq \frac{1}{\sqrt{2}} \left\| \left((\chi(t_i^n) - \chi(t_i^{-n}))v_2 - (S_2, S_3)(t_i^n \cdot x_0 - t_i^{-n} \cdot x_0) \right)_{i \in \{1,\ldots,d_2\}} \right\|
\end{aligned}
$$
$$(3.21)$$

for all $n \in \{1, \ldots, N_0\}$. For all $j \in \{1, \ldots, d_2\}$ we define $\alpha_j \in [0, 2\pi)$ by $e^{i\alpha_j} = \chi(t_j)$. Let $x_{0,1} \in \mathbb{R}^{d_1}$ and $x_{0,2} \in \mathbb{R}^{d_2}$ be such that $x_0 = \begin{pmatrix} x_{0,1} \\ x_{0,2} \end{pmatrix}$. For all $j \in \{1, \ldots, q\}$ we define $n_j = d_1 - 2(q - j + 1)$, $m_j = 2(q - j)$ and

$$
b_j = S_2(0_{n_j, n_j} \oplus \left(\begin{smallmatrix} 0 & -2 \\ 2 & 0 \end{smallmatrix} \right) \oplus 0_{m_j, m_j})x_{0,1} \in \mathbb{C}^{d_2}.
$$

Let $\tau_2 \colon \mathcal{T}^m \to \mathbb{R}^{d_2}$ be uniquely defined by the condition $\tau(t) = \begin{pmatrix} 0_{d_1} \\ \tau_2(t) \end{pmatrix}$ for all $t \in \mathcal{T}^m$. Then for all $i \in \{1, \ldots, d_2\}$ and $n \in \{1, \ldots, N_0\}$ we have

$$
\begin{aligned}
&(S_2, S_3)(t_i^n \cdot x_0 - t_i^{-n} \cdot x_0) \\
&= S_2\big(0_{d_1-2q} \oplus (R(n\theta_{i,1}) - R(-n\theta_{i,1})) \oplus \cdots \oplus (R(n\theta_{i,q}) - R(n\theta_{i,q}))\big)x_{0,1} \\
&\quad + 2nS_3\tau_2(t_i) \\
&= \sum_{j=1}^{q} \sin(n\theta_{i,j}) S_2\big(0_{n_j, n_j} \oplus \left(\begin{smallmatrix} 0 & -2 \\ 2 & 0 \end{smallmatrix} \right) \oplus 0_{m_j, m_j}\big)x_{0,1} + 2nS_3\tau_2(t_i) \\
&= \sum_{j=1}^{q} \sin(n\theta_{i,j}) b_j + 2nS_3\tau_2(t_i).
\end{aligned}
$$

For all $i \in \{1, \ldots, d_2\}$ and $j \in \{1, \ldots, q\}$ we define $B_{i,j} = b_j \otimes e_i^{\mathsf{T}} \in \mathbb{C}^{d_2 \times d_2}$. Let $T = 2\big(\tau_2(t_1), \ldots, \tau_2(t_{d_2})\big) \in \text{GL}(d_2)$. By equation (3.21) for all $n \in \{1, \dots, N_0\}$ we have

$$\text{LHS} \geq \frac{1}{\sqrt{2}} \left\| 2\mathrm{i}v_2 \otimes (\sin(n\alpha_1), \dots, \sin(n\alpha_{d_2})) \right.$$
$$\left. - \sum_{i=1}^{d_2} \sum_{j=1}^{q} \sin(n\theta_{i,j}) B_{i,j} - nS_3 T \right\|$$
$$\geq c_1 \left\| (2\mathrm{i}T^\mathsf{T} v_2) \otimes (\sin(n\alpha_1), \dots, \sin(n\alpha_{d_2})) \right.$$
$$\left. - \sum_{i=1}^{d_2} \sum_{j=1}^{q} \sin(n\theta_{i,j}) T^\mathsf{T} B_{i,j} - nT^\mathsf{T} S_3 T \right\|,$$

where $c_1 = \sigma_{\min}(T^{-\mathsf{T}})/\sqrt{2} > 0$, $\sigma_{\min}(M)$ denotes the minimum singular value of a matrix M and we used Corollary 9.6.7 in [10] in the last step. With equation (3.19) it follows

$$\text{LHS} \geq c_1 \|T^\mathsf{T} S_3 T\| \geq c_0 \|S_3\|,$$

where $c_0 = \sigma_{\min}(T)^2 c_1 > 0$.

By Proposition 3.20 it suffices to show that there exists a constant $c > 0$ such that $\| \cdot \|_{\mathcal{R}, \nabla} \geq c \| \cdot \|_{\mathcal{R}, \nabla, 0}$. Let $u \in U_{\text{per}}$. Let $N \in M_0$ be such that u is $\mathcal{T}^N$-periodic. In particular, m divides N. Let $v \colon \mathcal{G} \to \text{Skew}(d)$ be $\mathcal{T}^N$-periodic such that $\pi_{U_{\text{rot}}(\mathcal{R})}(\nabla_{\mathcal{R}} u(g)) = \nabla_{\mathcal{R}} u(g) - g_{\mathcal{R}} \circ v(g)$ for all $g \in \mathcal{G}$. Let

$$v_1 \colon \mathcal{G} \to \left\{ \begin{pmatrix} S_1 & S_2 \\ -S_2^\mathsf{T} & 0 \end{pmatrix} \,\middle|\, S_1 \in \text{Skew}(d_1), S_2 \in \mathbb{R}^{d_1 \times d_2} \right\}$$

and

$$v_2 \colon \mathcal{G} \to \{0_{d_1, d_1} \oplus S \mid S \in \text{Skew}(d_2)\}$$

such that $v = v_1 + v_2$. For all $g \in \mathcal{C}_m$ we define the functions

$$u_g \colon \mathcal{T}^m \to \mathbb{C}^d, \ t \mapsto u(gt)$$
$$v_g \colon \mathcal{T}^m \to \text{Skew}(d, \mathbb{C}), \ t \mapsto v(gt)$$
$$v_{1,g} \colon \mathcal{T}^m \to \text{Skew}(d, \mathbb{C}), \ t \mapsto v_1(gt)$$

and

$$v_{2,g} \colon \mathcal{T}^m \to \text{Skew}(d, \mathbb{C}), \ t \mapsto v_2(gt).$$

Let $\mathcal{E}' = \{\chi \in \widehat{\mathcal{T}^m} \mid \chi \text{ is periodic}\}$. For all $g \in \mathcal{C}_m$ and $\chi \in \mathcal{E}'$ it holds

$$\widehat{v_g}(\chi) = \widehat{v_{1,g}}(\chi) + \widehat{v_{2,g}}(\chi),$$
$$\widehat{v_{1,g}}(\chi) \in \left\{ \begin{pmatrix} S_1 & S_2 \\ -S_2^\mathsf{T} & 0 \end{pmatrix} \,\middle|\, S_1 \in \text{Skew}(d_1, \mathbb{C}), S_2 \in \mathbb{C}^{d_1 \times d_2} \right\}$$

and
$$\widehat{v_{2,g}}(\chi) \in \{0_{d_1,d_1} \oplus S \mid S \in \mathrm{Skew}(d_2,\mathbb{C})\}.$$

We have

$$
\|u\|_{\mathcal{R},\nabla}^2 = \frac{1}{|\mathcal{C}_N|} \sum_{(g,t)\in\mathcal{C}_m\times(\mathcal{T}^m\cap\mathcal{C}_N)} \|\pi_{U_{\mathrm{rot}}(\mathcal{R})}(\nabla_{\mathcal{R}}u(gt))\|^2
$$

$$
= \frac{1}{|\mathcal{C}_N|} \sum_{g\in\mathcal{C}_m} \sum_{t\in\mathcal{T}^m\cap\mathcal{C}_N} \|\nabla_{\mathcal{R}}u(gt) - g_{\mathcal{R}}\circ v(gt)\|^2
$$

$$
\geq \frac{1}{|\mathcal{C}_N|} \sum_{g\in\mathcal{C}_m} \sum_{t\in\mathcal{T}^m\cap\mathcal{C}_N} \|\nabla_{\mathcal{R}_0}u(gt) - g_{\mathcal{R}_0}\circ v(gt)\|^2
$$

$$
= \frac{1}{|\mathcal{C}_N|} \sum_{g\in\mathcal{C}_m} \sum_{t\in\mathcal{T}^m\cap\mathcal{C}_N} \left\|\left(u_g(th) - L(h)^{\mathsf{T}}u_g(t)\right)_{h\in\mathcal{R}_0}\right.
$$
$$
\left. - g_{\mathcal{R}_0}\circ v_g(t)\right\|^2
$$

$$
= \frac{1}{|\mathcal{C}_N|} \sum_{g\in\mathcal{C}_m} |\mathcal{T}^m\cap\mathcal{C}_N| \sum_{\chi\in\mathcal{E}'} \left\|\left(\chi(h)^{-1}\widehat{u_g}(\chi) - L(h)^{\mathsf{T}}\widehat{u_g}(\chi)\right)_{h\in\mathcal{R}_0}\right.
$$
$$
\left. - g_{\mathcal{R}_0}\circ\widehat{v_g}(\chi)\right\|^2
$$

$$
\geq \frac{c_0^2}{|\mathcal{C}_N|} \sum_{g\in\mathcal{C}_m} |\mathcal{T}^m\cap\mathcal{C}_N| \sum_{\chi\in\mathcal{E}'} \|\widehat{v_{2,g}}(\chi)\|^2
$$

$$
= \frac{c_0^2}{|\mathcal{C}_N|} \sum_{g\in\mathcal{C}_m} \sum_{t\in\mathcal{T}^m\cap\mathcal{C}_N} \|v_{2,g}(t)\|^2
$$

$$
= \frac{c_0^2}{|\mathcal{C}_N|} \sum_{(g,t)\in\mathcal{C}_m\times(\mathcal{T}^m\cap\mathcal{C}_N)} \|v_2(gt)\|^2
$$

$$
= c_0^2\|v_2\|_2^2. \tag{3.22}
$$

In the first and last step we used that $\bigcup_{(g,t)\in\mathcal{C}_m\times(\mathcal{T}^m\cap\mathcal{C}_N)}\{gt\}$ is a representation set of $\mathcal{G}/\mathcal{T}^N$. In the fifth and seventh step we used Proposition 2.56 for the group $\mathcal{T}^m$ and $\mathcal{T}^N$-periodic functions and Lemma 2.58. Note that $\mathcal{T}^m\cap\mathcal{C}_N$ is a representation set of $\mathcal{T}^m/\mathcal{T}^N$. In the sixth step we used (3.20). Let $C = |\mathcal{R}|\max\{\|h\cdot x_0 - x_0\| \mid h\in\mathcal{R}\}$. We have

$$
\|u\|_{\mathcal{R},\nabla}^2 = \frac{1}{|\mathcal{C}_N|} \sum_{g\in\mathcal{C}_N} \|\nabla_{\mathcal{R}}u(g) - g_{\mathcal{R}}\circ v(g)\|^2
$$

$$
\geq \frac{1}{|\mathcal{C}_N|} \sum_{g\in\mathcal{C}_N} \left(\frac{1}{2}\|\nabla_{\mathcal{R}}u(g) - g_{\mathcal{R}}\circ v_1(g)\|^2 - \|g_{\mathcal{R}}\circ v_2(g)\|^2\right)
$$

$$\geq \frac{1}{|\mathcal{C}_N|} \sum_{g \in \mathcal{C}_N} \left(\frac{1}{2} \|\pi_{U_{\mathrm{rot},0}(\mathcal{R})}(\nabla_{\mathcal{R}} u(g)))\|^2 - C\|v_2(g)\|^2 \right)$$

$$= \frac{1}{2}\|u\|_{\mathcal{R},\nabla,0}^2 - C\|v_2\|_2^2, \tag{3.23}$$

where in the second step we used that $(a-b)^2 \geq a^2/2 - b^2$ for all $a, b \geq 0$. Let $c_2 = \min\{1/2, c_0^2/(2C)\}$. By (3.22) and (3.23) we have

$$\|u\|_{\mathcal{R},\nabla}^2 \geq \frac{1}{2}\|u\|_{\mathcal{R},\nabla}^2 + c_2\|u\|_{\mathcal{R},\nabla}^2$$

$$\geq \frac{c_0^2}{2}\|v_2\|_2^2 + c_2\left(\frac{1}{2}\|u\|_{\mathcal{R},\nabla,0}^2 - C\|v_2\|_2^2\right)$$

$$\geq \frac{c_2}{2}\|u\|_{\mathcal{R},\nabla,0}^2.$$

Thus, we have $\|\cdot\|_{\mathcal{R},\nabla} \geq \sqrt{c_2/2}\|\cdot\|_{\mathcal{R},\nabla,0}$. □

3.2.4. The kernel of the seminorm $\|\cdot\|_{\mathcal{R}}$

In this section we define and analyze the vector spaces U_{trans} and $U_{\mathrm{rot},0,0}$ which correspond in the physical model to the space of all translations and infinitesimal rotations about the subspace $\{0_{d_1}\} \times \mathbb{R}^{d_2}$ of $\mathcal{G} \cdot x_0$, respectively.

Definition 3.25. For all $\mathcal{R} \subset \mathcal{G}$ we define the vector spaces

$$U_{\mathrm{rot},0,0}(\mathcal{R}) := \Big\{ u: \mathcal{R} \to \mathbb{R}^d \,\Big|\, \exists S \in \mathrm{Skew}(d_1) \,\forall g \in \mathcal{R} : L(g)u(g)$$

$$= (S \oplus 0_{d_2,d_2})(g \cdot x_0 - x_0) \Big\}$$

$$\subset U_{\mathrm{rot},0}(\mathcal{R}) \cap \big\{ u: \mathcal{G} \to \mathbb{R}^{d_1} \times \{0_{d_2}\} \big\}$$

and

$$U_{\mathrm{iso},0,0}(\mathcal{R}) := U_{\mathrm{trans}}(\mathcal{R}) + U_{\mathrm{rot},0}(\mathcal{R}) \subset U_{\mathrm{iso},0}(\mathcal{R}).$$

For brevity, we define

$$U_{\mathrm{trans}} := U_{\mathrm{trans}}(\mathcal{G})$$
$$U_{\mathrm{rot},0,0} := U_{\mathrm{rot},0,0}(\mathcal{G})$$

and

$$U_{\mathrm{iso},0,0} := U_{\mathrm{iso},0,0}(\mathcal{G}).$$

Remark 3.26. We have $U_{\text{rot},0,0} \subset U_{\text{rot},0}(\mathcal{G})$. If $d \geq 2$ and $d_2 \geq 1$, then we have $U_{\text{rot},0,0} \subsetneqq U_{\text{rot},0}(\mathcal{G})$. Moreover, in general we have $U_{\text{trans}} \not\subset U_{\text{per}}$ and $U_{\text{rot},0,0} \not\subset U_{\text{per}}$. For example let $\alpha \in \mathbb{R} \setminus (2\pi\mathbb{Q})$ be an angle, $\mathcal{G} = \langle R(\alpha) \oplus (I_1, 1) \rangle < \mathrm{E}(3)$ and $x_0 = e_1$. Then we have $\dim(U_{\text{rot},0}(\mathcal{G})) = 3$, $\dim(U_{\text{rot},0,0}) = 1$ and $\dim(U_{\text{rot},0,0} \cap U_{\text{per}}) = 0$. Moreover, we have $\dim(U_{\text{trans}}) = 3$ and $\dim(U_{\text{trans}} \cap U_{\text{per}}) = 1$.

Example 3.27. If $d_1 = 1$ or $d_{\text{aff}} = d_2$, then we have $U_{\text{rot},0,0} = \{0\}$. In particular, if $\mathcal{G}$ is a space group, then we have $U_{\text{rot},0,0} = \{0\}$.

The next proposition characterizes the vector spaces $U_{\text{trans}}(\mathcal{R})$, $U_{\text{rot}}(\mathcal{R})$, $U_{\text{rot},0}(\mathcal{R})$, $U_{\text{rot},0,0}(\mathcal{R})$, $U_{\text{iso}}(\mathcal{R})$, $U_{\text{iso},0}(\mathcal{R})$ and $U_{\text{iso},0,0}(\mathcal{R})$ for appropriate $\mathcal{R} \subset \mathcal{G}$. In particular, the proposition characterizes U_{trans}, $U_{\text{rot},0,0}$ and $U_{\text{iso},0,0}$ since $\mathcal{G}$ has a subset with Property 1.

Proposition 3.28. *Suppose that* $\mathcal{R} \subset \mathcal{G}$ *has a subset with Property* 1. *Then the maps*

$$\varphi_1 \colon \mathbb{R}^d \to U_{\text{trans}}(\mathcal{R})$$
$$a \mapsto \big(\mathcal{R} \to \mathbb{R}^d, g \mapsto L(g)^\mathsf{T} a\big),$$

$$\varphi_2 \colon \mathbb{R}^{d_3 \times d_{\text{aff}}} \times \text{Skew}(d_{\text{aff}}) \to U_{\text{rot}}(\mathcal{R})$$
$$(A_1, A_2) \mapsto \left(\mathcal{R} \to \mathbb{R}^d, g \mapsto L(g)^\mathsf{T} \begin{pmatrix} 0 & A_1 \\ -A_1^\mathsf{T} & A_2 \end{pmatrix} (g \cdot x_0 - x_0)\right),$$

$$\varphi_3 \colon \mathbb{R}^{d_3 \times d_4} \times \mathbb{R}^{d_3 \times d_2} \times \text{Skew}(d_4) \times \mathbb{R}^{d_4 \times d_2} \to U_{\text{rot},0}(\mathcal{R})$$
$$(A_1, A_2, A_3, A_4)$$
$$\mapsto \left(\mathcal{R} \to \mathbb{R}^d, g \mapsto L(g)^\mathsf{T} \begin{pmatrix} 0 & A_1 & A_2 \\ -A_1^\mathsf{T} & A_3 & A_4 \\ -A_2^\mathsf{T} & -A_4^\mathsf{T} & 0 \end{pmatrix} (g \cdot x_0 - x_0)\right),$$

and

$$\varphi_4 \colon \mathbb{R}^{d_3 \times d_4} \times \text{Skew}(d_4) \to U_{\text{rot},0,0}(\mathcal{R})$$
$$(A_1, A_2) \mapsto \left(\mathcal{R} \to \mathbb{R}^d, g \mapsto L(g)^\mathsf{T} \left(\begin{pmatrix} 0 & A_1 \\ -A_1^\mathsf{T} & A_2 \end{pmatrix} \oplus 0_{d_2, d_2}\right) (g \cdot x_0 - x_0)\right)$$

are isomorphisms, where $d_3 = d - d_{\text{aff}}$ *and* $d_4 = d_{\text{aff}} - d_2$. *In particular, we have*

$$\dim(U_{\text{trans}}(\mathcal{R})) = d$$
$$\dim(U_{\text{rot}}(\mathcal{R})) = d_{\text{aff}}(d - \tfrac{1}{2}d_{\text{aff}} - \tfrac{1}{2}),$$
$$\dim(U_{\text{rot},0}(\mathcal{R})) = d_3 d_{\text{aff}} + \tfrac{1}{2}d_4(d_{\text{aff}} + d_2 - 1)$$

and

$$\dim(U_{\text{rot},0,0}(\mathcal{R})) = d_4(d_3 + d_1 - 1)/2.$$

Moreover we have

$$U_{\text{iso}}(\mathcal{R}) = U_{\text{trans}}(\mathcal{R}) \oplus U_{\text{rot}}(\mathcal{R}),$$
$$U_{\text{iso},0}(\mathcal{R}) = U_{\text{trans}}(\mathcal{R}) \oplus U_{\text{rot},0}(\mathcal{R})$$

and

$$U_{\text{iso},0,0}(\mathcal{R}) = U_{\text{trans}}(\mathcal{R}) \oplus U_{\text{rot},0,0}(\mathcal{R}).$$

Proof. Let $d_3 = d - d_{\text{aff}}$, $d_4 = d_{\text{aff}} - d_2$ and $\mathcal{R} \subset \mathcal{G}$ be such that $\mathcal{R}$ has a subset with Property 1. In particular, we have $id \in \mathcal{R}$.

Since $L(id) = I_d$, the map φ_1 is injective and thus, an isomorphism.

Now we prove that φ_3 is an isomorphism. The map φ_3 is well-defined and linear.

First we show that φ_3 is surjective. Let $u \in U_{\text{rot},0}(\mathcal{R})$. There exist some $A_1 \in \text{Skew}(d_1)$ and $A_2 \in \mathbb{R}^{d_1 \times d_2}$ such that

$$L(g)u(g) = \begin{pmatrix} A_1 & A_2 \\ -A_2^\mathsf{T} & 0 \end{pmatrix}(g \cdot x_0 - x_0) \quad \text{for all } g \in \mathcal{G}.$$

Let $A_3 \in \text{Skew}(d_3)$, $A_4 \in \mathbb{R}^{d_3 \times d_4}$, $A_5 \in \text{Skew}(d_4)$, $A_6 \in \mathbb{R}^{d_3 \times d_2}$ and $A_7 \in \mathbb{R}^{d_4 \times d_2}$ be such that

$$A_1 = \begin{pmatrix} A_3 & A_4 \\ -A_4^\mathsf{T} & A_5 \end{pmatrix} \text{ and } A_2 = \begin{pmatrix} A_6 \\ A_7 \end{pmatrix}.$$

Since $\mathcal{G} \cdot x_0 \subset \{0_{d_3}\} \times \mathbb{R}^{d_{\text{aff}}}$, we have $\varphi_3(A_4, A_6, A_5, A_7) = u$.

Now we show that φ_3 is injective. Let the matrices $A_1, B_1 \in \mathbb{R}^{d_3 \times d_4}$, $A_2, B_2 \in \mathbb{R}^{d_3 \times d_2}$, $A_3, B_3 \in \text{Skew}(d_4)$ and $A_4, B_4 \in \mathbb{R}^{d_4 \times d_2}$ be such that $\varphi_3(A_1, A_2, A_3, A_4) = \varphi_3(B_1, B_2, B_3, B_4)$. Let $\mathcal{R}' \subset \mathcal{R}$ be such that $\mathcal{R}'$ has Property 1. By Lemma 3.6 there exists some $C \in \mathbb{R}^{d_{\text{aff}} \times |\mathcal{R}'|}$ of rank d_{aff} such that

$$(g \cdot x_0 - x_0)_{g \in \mathcal{R}'} = \begin{pmatrix} 0 \\ C \end{pmatrix}.$$

The identity $\varphi_3(A_1, A_2, A_3, A_4) = \varphi_3(B_1, B_2, B_3, B_4)$ implies

$$\begin{pmatrix} 0 & A_1 & A_2 \\ -A_1^\mathsf{T} & A_3 & A_4 \\ -A_2^\mathsf{T} & -A_4^\mathsf{T} & 0 \end{pmatrix}(g \cdot x_0 - x_0) = \begin{pmatrix} 0 & B_1 & B_2 \\ -B_1^\mathsf{T} & B_3 & B_4 \\ -B_2^\mathsf{T} & -B_4^\mathsf{T} & 0 \end{pmatrix}(g \cdot x_0 - x_0)$$

for all $g \in \mathcal{R}$ and in particular, we have

$$\begin{pmatrix} (A_1 \ A_2)C \\ (A_3 \ A_4)C \end{pmatrix} = \begin{pmatrix} (B_1 \ B_2)C \\ (B_3 \ B_4)C \end{pmatrix}.$$

Since the rank of C is equal to the number of its rows, we have $A_i = B_i$ for all $i \in \{1, \ldots, 4\}$.

The proofs that φ_2 and φ_4 are isomorphisms are analogous.
For all $u \in U_{\mathrm{rot}}(\mathcal{R})$ we have $u(id) = 0$ and for all $u \in U_{\mathrm{trans}}(\mathcal{R})$ and $g \in \mathcal{R}$
we have $L(g)u(g) = u(id)$. This implies $U_{\mathrm{trans}}(\mathcal{R}) \cap U_{\mathrm{rot}}(\mathcal{R}) = \{0\}$ and
thus $U_{\mathrm{iso}}(\mathcal{R}) = U_{\mathrm{trans}}(\mathcal{R}) \oplus U_{\mathrm{rot}}(\mathcal{R})$. Analogously, we have $U_{\mathrm{iso},0}(\mathcal{R}) =$
$U_{\mathrm{trans}}(\mathcal{R}) \oplus U_{\mathrm{rot},0}(\mathcal{R})$ and $U_{\mathrm{iso},0,0}(\mathcal{R}) = U_{\mathrm{trans}}(\mathcal{R}) \oplus U_{\mathrm{rot},0,0}(\mathcal{R})$. $\qquad\square$

Lemma 3.29. *If the group $L(\mathcal{G})$ is finite, then we have $U_{\mathrm{iso},0,0} \subset U_{\mathrm{per}}$.*

Proof. Suppose that $L(\mathcal{G})$ is finite. Let $n = |L(\mathcal{G})|$. For all $g \in \mathcal{G}$ we have

$$L(g)^n = I_d. \tag{3.24}$$

By Theorem 2.17 there exists some $N \in M_0$ such that n divides N. Let
$u \in U_{\mathrm{iso},0,0}$. By definition there exist some $a \in \mathbb{R}^d$ and $S \in \mathrm{Skew}(d_1)$
such that

$$L(g)u(g) = a + (S \oplus 0)(g \cdot x_0) \text{ for all } g \in \mathcal{G}.$$

For all $g \in \mathcal{G}$ and $t \in \mathcal{T}$ we have

$$\begin{aligned}
u(gt^N) &= L(gt^N)^{-1}\big(a + (S \oplus 0)((gt^N) \cdot x_0 - x_0)\big) \\
&= L(t)^{-N} L(g)^{-1}\big(a + (S \oplus 0)(g \cdot (L(t)^N x_0) - x_0) \\
&\quad + (S \oplus 0)L(g)\tau(t^N)\big) \\
&= L(g)^{-1}\big(a + (S \oplus 0)(g \cdot x_0 - x_0)\big) \\
&= u(g),
\end{aligned}$$

where we used (3.24), that $L(\mathcal{G}) < \oplus(\mathrm{O}(d_1) \times \mathrm{O}(d_2))$ and that $\tau(\mathcal{G}) \subset$
$\{0_{d_1}\} \times \mathbb{R}^{d_2}$ in the second to last step. Thus, u is $\mathcal{T}^N$-periodic and we
have $u \in U_{\mathrm{per}}$. $\qquad\square$

The following theorem characterizes the kernel of the seminorm $\|\cdot\|_{\mathcal{R}}$.

Theorem 3.30. *Suppose that $\mathcal{R} \subset \mathcal{G}$ has Property 2. Then we have*

$$\ker(\|\cdot\|_{\mathcal{R}}) = U_{\mathrm{iso},0,0} \cap U_{\mathrm{per}}.$$

Proof. Suppose that $\mathcal{R} \subset \mathcal{G}$ has Property 2.
First we show that $U_{\mathrm{iso},0,0} \cap U_{\mathrm{per}} \subset \ker(\|\cdot\|_{\mathcal{R}})$:
Let $u \in U_{\mathrm{iso},0,0} \cap U_{\mathrm{per}}$. There exist some $a \in \mathbb{R}^d$ and $S \in \mathrm{Skew}(d)$ such
that

$$L(g)u(g) = a + S(g \cdot x_0 - x_0) \qquad \text{for all } g \in \mathcal{G}.$$

Let $g \in \mathcal{G}$. For all $h \in \mathcal{R}$ it holds

$$\begin{aligned}
L(h)u(gh) &= L(g)^{\mathsf{T}}a + L(g)^{\mathsf{T}}S((gh) \cdot x_0 - x_0) \\
&= L(g)^{\mathsf{T}}a + L(g)^{\mathsf{T}}S\tau(g) + L(g)^{\mathsf{T}}SL(g)(h \cdot x_0 - x_0).
\end{aligned}$$

Since $L(g)^{\mathsf{T}} SL(g) \in \mathrm{Skew}(d)$, we have $u(g \cdot)|_{\mathcal{R}} \in U_{\mathrm{iso}}(\mathcal{R})$.
Let $N \in M_0$ be such that u is $\mathcal{T}^N$-periodic. Since $g \in \mathcal{G}$ was arbitrary, we have

$$\|u\|_{\mathcal{R}}^2 = \frac{1}{|\mathcal{C}_N|} \sum_{g \in \mathcal{C}_N} \|\pi_{U_{\mathrm{iso}}(\mathcal{R})}(u(g \cdot)|_{\mathcal{R}})\|^2 = 0.$$

Thus, we have $u \in \ker(\| \cdot \|_{\mathcal{R}})$.
Now we show that $\ker(\| \cdot \|_{\mathcal{R}}) \subset U_{\mathrm{iso},0,0} \cap U_{\mathrm{per}}$:
Let $u \in \ker(\| \cdot \|_{\mathcal{R}})$. By definition of $\| \cdot \|_{\mathcal{R}}$ we have $u \in U_{\mathrm{per}}$. Let $g \in \mathcal{G}$. By Theorem 3.13 we have $u \in \ker(\| \cdot \|_{\mathcal{R} \cup \{g\}})$ and thus $u|_{\mathcal{R} \cup \{g\}} \in U_{\mathrm{iso}}(\mathcal{R} \cup \{g\})$. There exist some $a \in \mathbb{R}^d$ and $S \in \mathrm{Skew}(d)$ such that

$$L(h)u(h) = a + S(h \cdot x_0 - x_0) \qquad \text{for all } h \in \mathcal{R} \cup \{g\}. \qquad (3.25)$$

Since $\mathcal{R}$ has Property 2, it holds $id \in \mathcal{R}$ and thus, $a = u(id)$. In particular, the vector a is independent of g.
Since $\mathcal{R}$ has Property 2, by Lemma 3.6 there exists some $A \in \mathbb{R}^{d_{\mathrm{aff}} \times |\mathcal{R}|}$ of rank d_{aff} such that

$$(g \cdot x_0 - x_0)_{g \in \mathcal{R}} = \begin{pmatrix} 0_{d-d_{\mathrm{aff}},|\mathcal{R}|} \\ A \end{pmatrix}.$$

Since $\mathcal{G} \cdot x_0 \subset \{0_{d-d_{\mathrm{aff}}}\} \times \mathbb{R}^{d_{\mathrm{aff}}}$, without loss of generality we may assume that

$$S = \begin{pmatrix} 0 & S_1 \\ -S_1^{\mathsf{T}} & S_2 \end{pmatrix}$$

for some $S_1 \in \mathbb{R}^{(d-d_{\mathrm{aff}}) \times d_{\mathrm{aff}}}$ and $S_2 \in \mathrm{Skew}(d_{\mathrm{aff}})$. By equation (3.25) we have

$$(L(h)u(h) - a)_{h \in \mathcal{R}} = \begin{pmatrix} 0 & S_1 \\ -S_1^{\mathsf{T}} & S_2 \end{pmatrix} \begin{pmatrix} 0 \\ A \end{pmatrix} = \begin{pmatrix} S_1 A \\ S_2 A \end{pmatrix}. \qquad (3.26)$$

Since the rank of A is equal to the number of its rows, by (3.26) the matrix S is independent of g.
Since $g \in \mathcal{G}$ was arbitrary, we have

$$L(g)u(g) = a + S(g \cdot x_0 - x_0) \qquad \text{for all } g \in \mathcal{G}. \qquad (3.27)$$

Let $C = \sup\{\|u(g)\| \mid g \in \mathcal{G}\}$. Since u is periodic, we have $C < \infty$. Let $t \in \mathcal{T}$. By (3.27) for all $n \in \mathbb{N}$ we have

$$n\|S\tau(t)\| = \|S\tau(t^n)\| = \|L(t^n)u(t^n) - a - SL(t^n)x_0 + Sx_0\|$$
$$\leq 2C + 2\|S\|\|x_0\|$$

and thus, $S\tau(t) = 0$. Since $t \in \mathcal{T}$ was arbitrary, we have

$$Sx = 0 \quad \text{for all } x \in \operatorname{span}(\{\tau(t) \mid t \in \mathcal{T}\}) = \{0_{d_1}\} \times \mathbb{R}^{d_2},$$

and thus, $S \in \oplus(\operatorname{Skew}(d_1) \times \{0_{d_2,d_2}\})$. By (3.27) we have $u \in U_{\mathrm{iso},0,0}$. □

Corollary 3.31. *Suppose that $L(\mathcal{G})$ is finite and $\mathcal{R} \subset \mathcal{G}$ has property 2. Then we have*

$$\ker(\|\cdot\|_{\mathcal{R}}) = U_{\mathrm{iso},0,0}.$$

Moreover, the map

$$\mathbb{R}^d \times \mathbb{R}^{d_3 \times d_4} \times \operatorname{Skew}(d_4) \to \ker(\|\cdot\|_{\mathcal{R}})$$
$$(a, A_1, A_2)$$
$$\mapsto \left(\mathcal{G} \to \mathbb{R}^d, g \mapsto L(g)^{\mathsf{T}}\left(a + \left(\left(\begin{smallmatrix} 0 & A_1 \\ -A_1^{\mathsf{T}} & A_2 \end{smallmatrix} \right) \oplus 0_{d_2,d_2} \right)(g \cdot x_0 - x_0) \right) \right)$$

is an isomorphism and in particular we have

$$\dim(\ker(\|\cdot\|_{\mathcal{R}})) = d + d_4(d_3 + d_1 - 1)/2,$$

where $d_3 = d - d_{\mathrm{aff}}$ and $d_4 = d_{\mathrm{aff}} - d_2$.

Proof. The assertion is clear by Theorem 3.30, Lemma 3.29 and Proposition 3.28. □

Corollary 3.32. *Suppose that $\mathcal{G}$ is a space group and $\mathcal{R} \subset \mathcal{G}$ has property 2. Then we have*

$$\ker(\|\cdot\|_{\mathcal{R}}) = U_{\mathrm{trans}}.$$

Proof. This is clear by Corollary 3.31 and Example 3.27. □

Example 3.33. We present an example which shows that in Theorem 3.13 the premise that $\mathcal{R}_1$ and $\mathcal{R}_2$ have Property 2 cannot be weakened to the premise that $\mathcal{R}_1$ and $\mathcal{R}_2$ are generating sets of $\mathcal{G}$ and have Property 1.
Suppose that $d = 2$, $d_1 = 1$, $d_2 = 1$, $t = (I_2, e_2)$, $\mathcal{G} = \langle t \rangle$, $x_0 = 0$, $\mathcal{R}_1 = \{id, t\}$ and $\mathcal{R}_2 = \{id, t, t^2\}$. The set $\mathcal{R}_1$ generates $\mathcal{G}$ and has Property 1 but does not have Property 2. The set $\mathcal{R}_2$ has Property 2. Using that the seminorms $\|\cdot\|_{\mathcal{R}}$ and $\|\cdot\|_{\mathcal{R}\setminus\{id\},\nabla}$ are equivalent by Proposition 3.20, it follows

$$\ker(\|\cdot\|_{\mathcal{R}_1}) = \{u \in U_{\mathrm{per}} \mid \exists a \in \mathbb{R} \, \forall g \in \mathcal{G} : u_2(g) = a\}.$$

By Corollary 3.31 and Example 3.27 we have

$$\ker(\|\cdot\|_{\mathcal{R}_2}) = U_{\mathrm{iso},0,0} = U_{\mathrm{trans}}.$$

Since the kernels of $\|\cdot\|_{\mathcal{R}_1}$ and $\|\cdot\|_{\mathcal{R}_2}$ are not equal, the seminorms $\|\cdot\|_{\mathcal{R}_1}$ and $\|\cdot\|_{\mathcal{R}_2}$ are not equivalent.

The following theorem summarizes the main results of this section.

Theorem 3.34. *Suppose that the sets $\mathcal{R}_1, \mathcal{R}_2 \subset \mathcal{G}$ have Property 2. Then the seminorms $\|\cdot\|_{\mathcal{R}_1}$, $\|\cdot\|_{\mathcal{R}_2}$, $\|\cdot\|_{\mathcal{R}_1,0}$, $\|\cdot\|_{\mathcal{R}_2,0}$, $\|\cdot\|_{\mathcal{R}_1,\nabla}$, $\|\cdot\|_{\mathcal{R}_2,\nabla}$, $\|\cdot\|_{\mathcal{R}_1,\nabla,0}$ and $\|\cdot\|_{\mathcal{R}_2,\nabla,0}$ are equivalent and their kernel is $U_{\mathrm{iso},0,0} \cap U_{\mathrm{per}}$.*

Proof. This is clear by Theorem 3.13, Proposition 3.20, Theorem 3.24 and Theorem 3.30. □

3.3. The seminorm $\|\cdot\|_{\mathcal{R},0,0}$

Recall Definition 3.25.

Definition 3.35. For all finite sets $\mathcal{R} \subset \mathcal{G}$ we define the seminorms

$$\|\cdot\|_{\mathcal{R},0,0}\colon U_{\mathrm{per}} \to [0,\infty)$$
$$u \mapsto \left(\frac{1}{|\mathcal{C}_N|} \sum_{g \in \mathcal{C}_N} \|\pi_{U_{\mathrm{iso},0,0}(\mathcal{R})}(u(g\cdot)|_{\mathcal{R}})\|^2\right)^{\frac{1}{2}}$$

and

$$\|\cdot\|_{\mathcal{R},\nabla,0,0}\colon U_{\mathrm{per}} \to [0,\infty)$$
$$u \mapsto \left(\frac{1}{|\mathcal{C}_N|} \sum_{g \in \mathcal{C}_N} \|\pi_{U_{\mathrm{rot},0,0}(\mathcal{R})}(\nabla_{\mathcal{R}} u(g))\|^2\right)^{\frac{1}{2}},$$

where u is $\mathcal{T}^N$-periodic and the maps $\pi_{U_{\mathrm{iso},0,0}(\mathcal{R})}$ and $\pi_{U_{\mathrm{rot},0,0}(\mathcal{R})}$ are the orthogonal projections on $\{u\colon \mathcal{R} \to \mathbb{R}^d\}$ with respect to the norm $\|\cdot\|$ with kernels $U_{\mathrm{iso},0,0}(\mathcal{R})$ and $U_{\mathrm{rot},0,0}(\mathcal{R})$, respectively.

Remark 3.36. For all finite sets $\mathcal{R} \subset \mathcal{G}$ we have $\|\cdot\|_{\mathcal{R}} \leq \|\cdot\|_{\mathcal{R},0,0}$, but the seminorms $\|\cdot\|_{\mathcal{R}}$ and $\|\cdot\|_{\mathcal{R},0,0}$ need not be equivalent, see Proposition 3.43.

Theorem 3.37. *Suppose that $\mathcal{R}_1, \mathcal{R}_2 \subset \mathcal{G}$ have Property 2. Then the seminorms $\|\cdot\|_{\mathcal{R}_1,0,0}$, $\|\cdot\|_{\mathcal{R}_2,0,0}$, $\|\cdot\|_{\mathcal{R}_1,\nabla,0,0}$, and $\|\cdot\|_{\mathcal{R}_2,\nabla,0,0}$ are equivalent and their kernel is $U_{\mathrm{iso},0,0} \cap U_{\mathrm{per}}$.*

Proof. Suppose that $\mathcal{R}_1, \mathcal{R}_2 \subset \mathcal{G}$ have Property 2. The proof that the seminorms $\|\cdot\|_{\mathcal{R}_1,0,0}$ and $\|\cdot\|_{\mathcal{R}_2,0,0}$ are equivalent is analogous to the proof of Theorem 3.13: For all finite sets we define the seminorm

$$p_{0,\mathcal{R}} : \{u \colon \mathcal{R} \to \mathbb{R}^d\} \to [0, \infty)$$
$$u \mapsto \|\pi_{U_{\mathrm{iso},0,0}(\mathcal{R})}(u)\|,$$

where $\pi_{U_{\mathrm{iso},0,0}}(\mathcal{R})$ is the orthogonal projection on $\{u \colon \mathcal{R} \to \mathbb{R}^d\}$ with respect to the norm $\|\cdot\|$ with kernel $U_{\mathrm{iso},0,0}(\mathcal{R})$. Analogously to Lemma 3.8 the kernel of the seminorm $p_{0,\mathcal{R}}$ is $U_{\mathrm{iso},0,0}(\mathcal{R})$. Analogously to Definition 3.9 for all finite sets $\mathcal{R}_1, \mathcal{R}_2, \mathcal{R}_3 \subset \mathcal{G}$ such that $\mathcal{R}_1 \subset \mathcal{R}_3\mathcal{R}_2$, we define the seminorm

$$q_{0,\mathcal{R}_1,\mathcal{R}_2,\mathcal{R}_3} : \{u \colon \mathcal{R}_1 \to \mathbb{R}^d\} \to [0, \infty)$$
$$u \mapsto \inf_{\substack{v \colon \mathcal{R}_3\mathcal{R}_2 \to \mathbb{R}^d \\ v|_{\mathcal{R}_1} = u}} \left(\sum_{g \in \mathcal{R}_3} p_{0,\mathcal{R}_2}^2 \big(v(g \cdot)|_{\mathcal{R}_2} \big) \right)^{\frac{1}{2}},$$

where the infimum is even a minimum, see Lemma 3.10. Analogously to Lemma 3.12 for all $\mathcal{R}_1, \mathcal{R}_2 \subset \mathcal{G}$ such that $\mathcal{R}_1$ has Property 1 and $\mathcal{R}_2$ has Property 2 we have that there exists a finite set $\mathcal{R}_3 \subset \mathcal{G}$ such that $\mathcal{R}_1 \subset \mathcal{R}_3\mathcal{R}_2$ and the seminorms $p_{\mathcal{R}_1}$ and $q_{\mathcal{R}_1,\mathcal{R}_2,\mathcal{R}_3}$ are equivalent. Analogously to the proof of Theorem 3.13 this implies that the seminorms $\|\cdot\|_{\mathcal{R}_1,0,0}$ and $\|\cdot\|_{\mathcal{R}_2,0,0}$ are equivalent.

Analogously to the proof of Proposition 3.20, the seminorms $\|\cdot\|_{\mathcal{R},0,0}$ and $\|\cdot\|_{\mathcal{R},\nabla,0,0}$ are equivalent for all finite sets $\mathcal{R} \subset \mathcal{G}$ such that $id \in \mathcal{R}$. In particular, if $\mathcal{R} \subset \mathcal{G}$ has Property 2, then $\|\cdot\|_{\mathcal{R},0,0}$ and $\|\cdot\|_{\mathcal{R},\nabla,0,0}$ are equivalent.

Suppose that $\mathcal{R} \subset \mathcal{G}$ has property 2. Analogously to the proof of Theorem 3.30, we have $U_{\mathrm{iso},0,0} \cap U_{\mathrm{per}} \subset \ker(\|\cdot\|_{\mathcal{R},0,0})$. Since $\|\cdot\|_{\mathcal{R}} \leq \|\cdot\|_{\mathcal{R},0,0}$, by Theorem 3.30 we have $\ker(\|\cdot\|_{\mathcal{R},0,0}) \subset U_{\mathrm{iso},0,0} \cap U_{\mathrm{per}}$. $\square$

3.4. The seminorm $\|\nabla_{\mathcal{R}} \cdot \|_2$

Definition 3.38. For all finite sets $\mathcal{R} \subset \mathcal{G}$ we define the norm

$$\|\cdot\|_2 : \{u \colon \mathcal{G} \to \{v \colon \mathcal{R} \to \mathbb{R}^d\} \mid u \text{ is periodic}\} \to [0, \infty)$$
$$u \mapsto \left(\frac{1}{|\mathcal{C}_N|} \sum_{g \in \mathcal{C}_N} \|u(g)\|^2 \right)^{\frac{1}{2}} \quad \text{if } u \text{ is } \mathcal{T}^N\text{-periodic.}$$

Remark 3.39. For all finite sets $\mathcal{R} \subset \mathcal{G}$ the map

$$\left(\{u \colon \mathcal{G} \to \{v \colon \mathcal{R} \to \mathbb{R}^d\} \mid u \text{ is periodic}\}, \|\cdot\|_2\right) \to \left(L^\infty_{\mathrm{per}}(\mathcal{G}, \mathbb{R}^{d \times |\mathcal{R}|}), \|\cdot\|_2\right)$$
$$u \mapsto (\mathcal{G} \to \mathbb{R}^{d \times |\mathcal{R}|}, g \mapsto (v(g))_{g \in \mathcal{R}})$$

is an isomorphism. Thus there is no ambiguity between the above definition and Definition 2.52.

Theorem 3.40. *Let $\mathcal{R}_1, \mathcal{R}_2 \subset \mathcal{G}$ be finite generating sets of $\mathcal{G}$. Then the seminorms $\|\nabla_{\mathcal{R}_1} \cdot \|_2$ and $\|\nabla_{\mathcal{R}_2} \cdot \|_2$ on U_{per} are equivalent and their kernel is $U_{\mathrm{trans}} \cap U_{\mathrm{per}}$.*

Proof. Let $\mathcal{R}_1, \mathcal{R}_2 \subset \mathcal{G}$ be finite generating sets of $\mathcal{G}$. Analogously to Lemma 3.3, the functions $\|\nabla_{\mathcal{R}_1} \cdot \|_2$ and $\|\nabla_{\mathcal{R}_2} \cdot \|_2$ are seminorms.
First we show that the seminorms $\|\nabla_{\mathcal{R}_1} \cdot \|_2$ and $\|\nabla_{\mathcal{R}_2} \cdot \|_2$ are equivalent. It suffices to show that there exists a constant $C > 0$ such that $\|\nabla_{\mathcal{R}_1} \cdot \|_2 \le C \|\nabla_{\mathcal{R}_2} \cdot \|_2$. Since $\mathcal{R}_2$ generates $\mathcal{G}$, for every $r \in \mathcal{R}_1$ there exist some $n_r \in \mathbb{N}$ and $s_{r,1}, \ldots, s_{r,n_r} \in \mathcal{R}_2 \cup \mathcal{R}_2^{-1}$ such that $r = s_{r,1} \ldots s_{r,n_r}$. Let $u \in U_{\mathrm{per}}$. Let $N \in M_0$ be such that u is $\mathcal{T}^N$-periodic. Then we have

$$\|\nabla_{\mathcal{R}_1} u\|_2^2 = \frac{1}{|\mathcal{C}_N|} \sum_{g \in \mathcal{C}_N} \|\nabla_{\mathcal{R}_1} u(g)\|^2$$

$$= \frac{1}{|\mathcal{C}_N|} \sum_{g \in \mathcal{C}_N} \sum_{r \in \mathcal{R}_1} \|L(r)u(gr) - u(g)\|^2$$

$$= \frac{1}{|\mathcal{C}_N|} \sum_{g \in \mathcal{C}_N} \sum_{r \in \mathcal{R}_1} \left\| \sum_{i=1}^{n_r} L(s_{r,1} \ldots s_{r,i-1})\big(L(s_{r,i})u(gs_{r,1} \ldots s_{r,i})\right.$$
$$\left. - u(gs_{r,1} \ldots s_{r,i-1})\big) \right\|^2$$

$$\le \frac{1}{|\mathcal{C}_N|} \sum_{g \in \mathcal{C}_N} \sum_{r \in \mathcal{R}_1} \left(\sum_{i=1}^{n_r} \|L(s_{r,i})u(gs_{r,1} \ldots s_{r,i}) - u(gs_{r,1} \ldots s_{r,i-1})\| \right)^2$$

$$\le \frac{1}{|\mathcal{C}_N|} \sum_{g \in \mathcal{C}_N} \sum_{r \in \mathcal{R}_1} n_r \sum_{i=1}^{n_r} \|L(s_{r,i})u(gs_{r,1} \ldots s_{r,i}) - u(gs_{r,1} \ldots s_{r,i-1})\|^2$$

$$\le \frac{C}{|\mathcal{C}_N|} \sum_{\tilde{g} \in \mathcal{C}_N} \sum_{s \in \mathcal{R}_2} \|L(s)u(\tilde{g}s) - u(\tilde{g})\|^2$$

$$= C \|\nabla_{\mathcal{R}_2} u\|_2^2,$$

where $C = \sum_{r \in \mathcal{R}_1} n_r^2$. In the fifth step we used that the arithmetic mean is lower or equal than the root mean square. In the sixth step, if $s_{r,i} \in \mathcal{R}_2$, we substituted $gs_{r,1} \ldots s_{r,i-1}$ by $\tilde{g}$, and if $s_{r,i} \in \mathcal{R}_2^{-1}$, we substituted $gs_{r,1} \ldots s_{r,i}$ by $\tilde{g}$.

Let $\mathcal{R} = \mathcal{R}_1$. Now we show that $\ker(\|\nabla_{\mathcal{R}} \cdot \|_2) = U_{\text{trans}} \cap U_{\text{per}}$. It is clear that $U_{\text{trans}} \cap U_{\text{per}} \subset \ker(\|\nabla_{\mathcal{R}} \cdot \|_2)$. If $u \in \ker(\|\nabla_{\mathcal{R}} \cdot \|_2)$, then for all $g \in \mathcal{G}$ we have

$$0 = \|\nabla_{\mathcal{R} \cup \{g\}} u\|_2 \geq \|L(g)u(g) - u(id)\|, \tag{3.28}$$

where we used that the seminorms $\|\nabla_{\mathcal{R}} \cdot \|_2$ and $\|\nabla_{\mathcal{R} \cup \{g\}} \cdot \|_2$ are equivalent. By (3.28) we have $L(g)u(g) = u(id)$ for all $g \in \mathcal{G}$ and thus $u \in U_{\text{trans}}$. $\qquad \square$

Remark 3.41. For all finite sets $\mathcal{R} \subset \mathcal{G}$ we have $\| \cdot \|_{\mathcal{R},0,0} \leq \|\nabla_{\mathcal{R}} \cdot \|_2$, but the seminorms $\| \cdot \|_{\mathcal{R},0,0}$ and $\|\nabla_{\mathcal{R}} \cdot \|$ need not be equivalent since their kernels are not equal, see Theorem 3.37 and Theorem 3.40.

Theorem 3.24 yields the following corollary.

Corollary 3.42. *(A discrete Korn's inequality for space groups) Suppose that $\mathcal{G}$ is a space group and $\mathcal{R} \subset \mathcal{G}$ has Property 2. Then the seminorms $\| \cdot \|_{\mathcal{R}}$, $\| \cdot \|_{\mathcal{R},0,0}$ and $\|\nabla_{\mathcal{R}} \cdot \|_2$ are equivalent.*

Proof. Suppose that $\mathcal{G}$ is a space group and $\mathcal{R} \subset \mathcal{G}$ has Property 2. Then we have $U_{\text{rot},0}(\mathcal{R}) = U_{\text{rot},0,0}(\mathcal{R}) = \{0\}$ and $\| \cdot \|_{\mathcal{R},\nabla,0} = \| \cdot \|_{\mathcal{R},\nabla,0,0} = \|\nabla_{\mathcal{R}} \cdot \|_2$ With Theorem 3.34 and Theorem 3.37 follows the assertion. $\qquad \square$

3.5. Fourier transformation of a seminorm

Proposition 3.43 and Proposition 3.44 are very similar. In Proposition 3.43 we have $d_{\text{aff}} = 1$ and in Proposition 3.44 we have $d_{\text{aff}} = 2$.

Proposition 3.43. *Suppose that $t = (I_2, e_2) \in \text{E}(2)$, $\mathcal{G} = \langle t \rangle < \text{E}(2)$, $x_0 = 0 \in \mathbb{R}^2$ and $\mathcal{R} \subset \mathcal{G}$ has Property 2, e. g. $\mathcal{R} = \{id, t, t^2\}$. Then the seminorms $\| \cdot \|_{\mathcal{R},0,0}$ and $\|\nabla_{\mathcal{R}} \cdot \|_2$ are equivalent and there exist constants $C, c > 0$ such that for all $u \in U_{\text{per}}$ we have*

$$c\|\nabla_{\mathcal{R}} u\|_2^2 \leq \sum_{k \in [0,1) \cap \mathbb{Q}} |k|_1^2 \|\widehat{u}(\chi_k)\|^2 \leq C\|\nabla_{\mathcal{R}} u\|_2^2$$

and

$$c\|u\|_{\mathcal{R}}^2 \leq \sum_{k \in [0,1) \cap \mathbb{Q}} \left(|k|_1^4 |\widehat{u}_1(\chi_k)|^2 + |k|_1^2 |\widehat{u}_2(\chi_k)|^2 \right) \leq C\|u\|_{\mathcal{R}}^2,$$

where $| \cdot |_1 \colon \mathbb{R} \to [0, \infty)$, $k \mapsto \text{dist}(k, \mathbb{Z})$ is the distance to nearest integer function.

Proof. Suppose that $t = (I_2, e_2)$, $\mathcal{G} = \langle t \rangle$ and $x_0 = 0$. We have $d = 2$ and $d_1 = d_2 = 1$. The set $\{id, t, t^2\}$ has Property 2 and by Theorem 3.40 and Theorem 3.34 without loss of generality, let $\mathcal{R} = \{id, t, t^2\}$. Since $U_{\mathrm{rot},0,0}(\mathcal{R}) = \{0\}$, we have $\|\cdot\|_{\mathcal{R},\nabla,0,0} = \|\nabla_{\mathcal{R}} \cdot\|_2$ and thus the seminorms $\|\cdot\|_{\mathcal{R},0,0}$ and $\|\nabla_{\mathcal{R}} \cdot\|_2$ are equivalent by Theorem 3.37.

By Definition 2.29 for all $k \in \mathbb{R}$ and $n \in \mathbb{Z}$, we have $\chi_k(t^n) = e^{2\pi i n k}$. The maps

$$[0,1) \to \widehat{\mathcal{G}}, \ k \mapsto \chi_k$$

and

$$[0,1) \cap \mathbb{Q} \to \{\chi \in \widehat{\mathcal{G}} \mid \chi \text{ is periodic}\}, \ k \mapsto \chi_k$$

are bijective. Thus, without loss of generality, let $\mathcal{E} = \{\chi_k \mid k \in [0,1) \cap \mathbb{Q}\}$, see Definition 2.53.

Since $\{k \in [0,1) \mid e^{-2\pi i k} = 1\} = \{0\}$ and by Taylor's theorem, there exists a constant $c_T \in (0,1)$ such that for all $k \in [0,1)$ and $n \in \{1,2\}$ we have

$$c_T |k|_1 \le \left| e^{-2\pi i k} - 1 \right|, \tag{3.29}$$

$$c_T \left| e^{-2\pi i n k} - 1 \right| \le |k|_1, \tag{3.30}$$

and

$$c_T \left| e^{-2\pi i n k} - 1 + 2\pi i n k \right| \le |k|_1^2. \tag{3.31}$$

For all $u \in U_{\mathrm{per}}$ we have

$$\begin{aligned}
\|\nabla_{\mathcal{R}} u\|_2^2 &= \sum_{\chi \in \mathcal{E}} \left\| \widehat{\nabla_{\mathcal{R}} u}(\chi) \right\|^2 \\
&= \sum_{k \in [0,1) \cap \mathbb{Q}} \left\| (\chi_k(h)^{-1} \widehat{u}(\chi_k) - \widehat{u}(\chi_k))_{h \in \mathcal{R}} \right\|^2 \\
&= \sum_{k \in [0,1) \cap \mathbb{Q}} \sum_{n=1}^{2} \left| e^{-2\pi i n k} - 1 \right|^2 \left\| \widehat{u}(\chi_k) \right\|^2,
\end{aligned} \tag{3.32}$$

where we used Proposition 2.56 in the first step and Lemma 2.58 in the second step. Equations (3.29), (3.30) and (3.32) imply the first assertion. Now we show the second assertion. Let $\mathcal{R}' = \{t, t^2\}$. By Proposition 3.20 the seminorms $\|\cdot\|_{\mathcal{R}}$ and $\|\cdot\|_{\mathcal{R}',\nabla}$ are equivalent, i.e. there exist some constants $C, c > 0$ such that

$$c\|\cdot\|_{\mathcal{R}} \le \|\cdot\|_{\mathcal{R}',\nabla} \le C\|\cdot\|_{\mathcal{R}}. \tag{3.33}$$

We define the linear map

$$g_{\mathcal{R}'}\colon \operatorname{Skew}(2,\mathbb{C}) \to \mathbb{C}^{2 \times |\mathcal{R}'|}$$
$$S \mapsto \big(S(h \cdot x_0 - x_0)\big)_{h \in \mathcal{R}'}.$$

For all $u \in U_{\mathrm{per}}$ we have

$$\|u\|_{\mathcal{R}',\nabla}^2 = \inf\Big\{\|\nabla_{\mathcal{R}'}u - g_{\mathcal{R}'} \circ v\|_2^2 \,\Big|\, v \in L^{\infty}_{\mathrm{per}}(\mathcal{G}, \operatorname{Skew}(2,\mathbb{C}))\Big\}$$

$$= \inf\Big\{\sum_{\chi \in \mathcal{E}} \big\|\widehat{\nabla_{\mathcal{R}'}u}(\chi) - g_{\mathcal{R}'} \circ \tilde{v}(\chi)\big\|^2 \,\Big|\, \tilde{v} \in \bigoplus_{\chi \in \mathcal{E}} \operatorname{Skew}(2,\mathbb{C})\Big\}$$

$$= \sum_{\chi \in \mathcal{E}} \inf\Big\{\big\|\widehat{\nabla_{\mathcal{R}'}u}(\chi) - g_{\mathcal{R}'}(S)\big\|^2 \,\Big|\, S \in \operatorname{Skew}(2,\mathbb{C})\Big\}$$

$$= \sum_{k \in [0,1) \cap \mathbb{Q}} \inf\Big\{\big\|\big(\chi_k(h)^{-1}\widehat{u}(\chi_k) - \widehat{u}(\chi_k)$$

$$- \big(\begin{smallmatrix} 0 & -s \\ s & 0 \end{smallmatrix}\big)(h \cdot x_0 - x_0)\big)_{h \in \mathcal{R}'}\big\|^2 \,\Big|\, s \in \mathbb{C}\Big\}$$

$$= \sum_{k \in [0,1) \cap \mathbb{Q}} \inf\Big\{\sum_{n=1}^{2}\big\|(e^{-2\pi i n k} - 1)\widehat{u}(\chi_k) + nse_1\big\|^2 \,\Big|\, s \in \mathbb{C}\Big\},$$
(3.34)

where we used Proposition 2.56 in the second step and Lemma 2.58 in the fourth step.
It holds

$$\sum_{i=1}^{n} a_i^2 \leq \Big(\sum_{i=1}^{n} a_i\Big)^2 \leq n\sum_{i=1}^{n} a_i^2 \qquad (3.35)$$

for all $n \in \mathbb{N}$ and $a_1, \ldots, a_n \geq 0$.
We define the functions

$$f_1\colon [0,1) \times \mathbb{C}^2 \times \mathbb{C} \to [0,\infty)$$
$$(k,v,s) \mapsto \sum_{n=1}^{2}\big\|(e^{-2\pi i n k} - 1)v + nse_1\big\|$$

and

$$f_2\colon [0,1) \times \mathbb{C}^2 \to [0,\infty)$$
$$(k,v) \mapsto |k|_1^2|v_1| + |k|_1|v_2|.$$

By (3.33), (3.34) and (3.35) it suffices so show that there exist some constant $C, c > 0$ such that for all $(k, v) \in [0, 1) \times \mathbb{C}^2$ we have

$$c \inf_{s \in \mathbb{C}} f_1(k, v, s) \leq f_2(k, v) \leq C \inf_{s \in \mathbb{C}} f_1(k, v, s). \qquad (3.36)$$

First we show the left inequality of (3.36). By (3.30) and (3.31) for all $(k, v) \in [0, 1) \times \mathbb{C}^2$ we have

$$\inf_{s \in \mathbb{C}} f_1(k, v, s) \leq f_1(k, v, 2\pi i k v_1)$$

$$\leq \sum_{n=1}^{2} \left(\left| e^{-2\pi i n k} - 1 + 2\pi i n k \right| |v_1| + \left| e^{-2\pi i n k} - 1 \right| |v_2| \right)$$

$$\leq \tfrac{2}{c_T} f_2(k, v).$$

Now we show the right inequality of (3.36). Let $(k, v, s) \in [0, 1) \times \mathbb{C}^2 \times \mathbb{C}$. By (3.29) we have

$$f_1(k, v, s) \geq \left| e^{-2\pi i k} v_1 - v_1 + s \right| + \tfrac{1}{2} \left| e^{-4\pi i k} v_1 - v_1 + 2s \right|$$

$$\geq \tfrac{1}{2} \left| 2(e^{-2\pi i k} v_1 - v_1 + s) - (e^{-4\pi i k} v_1 - v_1 + 2s) \right|$$

$$= \tfrac{1}{2} \left| e^{-2\pi i k} - 1 \right|^2 |v_1|$$

$$\geq \tfrac{c_T^2}{2} |k|_1^2 |v_1| \qquad (3.37)$$

and

$$f_1(k, v, s) \geq \left| e^{-2\pi i k} - 1 \right| |v_2| \geq c_T |k|_1 |v_2|. \qquad (3.38)$$

By (3.37) and (3.38) we have

$$f_1(k, v, s) \geq \tfrac{c_T^2}{4} f_2(k, v). \qquad \square$$

Proposition 3.44. *Suppose that* $t = \left(\left(\begin{smallmatrix} -1 & 0 \\ 0 & 1 \end{smallmatrix} \right), e_2 \right) \in \mathrm{E}(2)$, $\mathcal{G} = \langle t \rangle <$ $\mathrm{E}(2)$, $x_0 = e_1 \in \mathbb{R}^2$ *and* $\mathcal{R} \subset \mathcal{G}$ *has Property 2, e.g.* $\mathcal{R} = \{ t^0, \ldots, t^3 \}$. *Then the seminorms* $\| \cdot \|_{\mathcal{R},0,0}$ *and* $\| \nabla_{\mathcal{R}} \cdot \|_2$ *are equivalent and there exist constants* $C, c > 0$ *such that for all* $u \in U_{\mathrm{per}}$ *we have*

$$c \| \nabla_{\mathcal{R}} u \|_2^2 \leq \sum_{k \in [0,1) \cap \mathbb{Q}} \left(|k - \tfrac{1}{2}|_1^2 |\widehat{u}_1(\chi_k)|^2 + |k|_1^2 |\widehat{u}_2(\chi_k)|^2 \right) \leq C \| \nabla_{\mathcal{R}} u \|_2^2$$

and

$$c \| u \|_{\mathcal{R}}^2 \leq \sum_{k \in [0,1) \cap \mathbb{Q}} \left(|k - \tfrac{1}{2}|_1^4 |\widehat{u}_1(\chi_k)|^2 + |k|_1^2 |2\pi i (k - \tfrac{1}{2}) \widehat{u}_1(\chi_k) - \widehat{u}_2(\chi_k)|^2 \right)$$

$$\leq C \| u \|_{\mathcal{R}}^2,$$

where $|\cdot|_1 \colon \mathbb{R} \to [0,\infty)$, $k \mapsto \mathrm{dist}(k,\mathbb{Z})$ *is the* distance to nearest integer function.

Proof. Suppose that $t = ((\begin{smallmatrix} -1 & 0 \\ 0 & 1 \end{smallmatrix}), e_2)$, $\mathcal{G} = \langle t \rangle$ and $x_0 = e_1$. We have $d = 2$ and $d_1 = d_2 = 1$. The set $\{t^0, \ldots, t^3\}$ has Property 2 and by Theorem 3.40 and Theorem 3.34 without loss of generality, let $\mathcal{R} = \{t^0, \ldots, t^3\}$. Since $U_{\mathrm{rot},0,0}(\mathcal{R}) = \{0\}$, we have $\|\cdot\|_{\mathcal{R},\nabla,0,0} = \|\nabla_{\mathcal{R}} \cdot\|_2$ and thus the seminorms $\|\cdot\|_{\mathcal{R},0,0}$ and $\|\nabla_{\mathcal{R}} \cdot\|_2$ are equivalent by Theorem 3.37.
By Definition 2.29 for all $k \in \mathbb{R}$ and $n \in \mathbb{Z}$, we have $\chi_k(t^n) = \mathrm{e}^{2\pi i n k}$. The maps

$$[0,1) \to \widehat{\mathcal{G}}, \ k \mapsto \chi_k$$

and

$$[0,1) \cap \mathbb{Q} \to \{\chi \in \widehat{\mathcal{G}} \,|\, \chi \text{ is periodic}\}, \ k \mapsto \chi_k$$

are bijective. Thus, without loss of generality, let $\mathcal{E} = \{\chi_k \,|\, k \in [0,1) \cap \mathbb{Q}\}$, see Definition 2.53.
Since $\{k \in [0,1) \,|\, \mathrm{e}^{-2\pi i k} = 1\} = \{0\}$, $\{k \in [0,1) \,|\, \mathrm{e}^{-2\pi i k} = -1\} = \{\tfrac{1}{2}\}$ and by Taylor's theorem, there exists a constant $c_T \in (0,1)$ such that for all $k \in [0,1)$ and $n \in \{1,2,3\}$ we have

$$c_T |k|_1 \le \left| \mathrm{e}^{-2\pi i k} - 1 \right|, \tag{3.39}$$

$$c_T \left| k - \tfrac{1}{2} \right|_1 \le \left| \mathrm{e}^{-2\pi i k} + 1 \right|, \tag{3.40}$$

$$c_T \left| \mathrm{e}^{-2\pi i n k} - 1 \right| \le |k|_1, \tag{3.41}$$

$$c_T \left| \mathrm{e}^{-2\pi i n k} - (-1)^n \right| \le \left| k - \tfrac{1}{2} \right|_1, \tag{3.42}$$

and

$$c_T \left| \mathrm{e}^{-2\pi i n k} - (-1)^n + (-1)^n 2\pi i n (k - \tfrac{1}{2}) \right| \le \left| k - \tfrac{1}{2} \right|_1^2. \tag{3.43}$$

For all $u \in U_{\mathrm{per}}$ we have

$$\|\nabla_{\mathcal{R}} u\|_2^2 = \sum_{\chi \in \mathcal{E}} \left\| \widehat{\nabla_{\mathcal{R}} u}(\chi) \right\|^2$$

$$= \sum_{k \in [0,1) \cap \mathbb{Q}} \left\| (\chi_k(h)^{-1} \widehat{u}(\chi_k) - L(h)^\mathsf{T} \widehat{u}(\chi_k))_{h \in \mathcal{R}} \right\|^2$$

$$= \sum_{k \in [0,1) \cap \mathbb{Q}} \sum_{n=1}^{3} \left\| \mathrm{e}^{-2\pi i n k} \widehat{u}(\chi_k) - \left(\begin{smallmatrix} -1 & 0 \\ 0 & 1 \end{smallmatrix}\right)^n \widehat{u}(\chi_k) \right\|^2$$

$$= \sum_{k\in[0,1)\cap\mathbb{Q}} \sum_{n=1}^{3} \Big(\big| e^{-2\pi i n k} - (-1)^n \big|^2 \big| \widehat{u}_1(\chi_k) \big|^2$$

$$+ \big| e^{-2\pi i n k} - 1 \big|^2 \big| \widehat{u}_2(\chi_k) \big|^2 \Big), \tag{3.44}$$

where we used Proposition 2.56 in the first step and Lemma 2.58 in the second step. Equations (3.39), (3.40), (3.41), (3.42) and (3.44) imply the first assertion.

Now we show the second assertion. Let $\mathcal{R}' = \{t^1, t^2, t^3\}$. By Proposition 3.20 the seminorms $\|\cdot\|_{\mathcal{R}}$ and $\|\cdot\|_{\mathcal{R}',\nabla}$ are equivalent, i.e. there exist some constants $C, c > 0$ such that

$$c\|\cdot\|_{\mathcal{R}} \leq \|\cdot\|_{\mathcal{R}',\nabla} \leq C\|\cdot\|_{\mathcal{R}}. \tag{3.45}$$

We define the linear map

$$g_{\mathcal{R}'} \colon \operatorname{Skew}(2,\mathbb{C}) \to \mathbb{C}^{2\times|\mathcal{R}'|}$$

$$S \mapsto \big(L(h)^{\mathsf{T}} S(h \cdot x_0 - x_0) \big)_{h\in\mathcal{R}'}.$$

For all $u \in U_{\mathrm{per}}$ we have

$$\|u\|_{\mathcal{R}',\nabla}^2 = \inf\Big\{ \|\nabla_{\mathcal{R}'} u - g_{\mathcal{R}'} \circ v\|_2^2 \,\Big|\, v \in L_{\mathrm{per}}^{\infty}(\mathcal{G}, \operatorname{Skew}(2,\mathbb{C})) \Big\}$$

$$= \inf\Big\{ \sum_{\chi\in\mathcal{E}} \big\| \widehat{\nabla_{\mathcal{R}'} u}(\chi) - g_{\mathcal{R}'} \circ \widetilde{v}(\chi) \big\|^2 \,\Big|\, \widetilde{v} \in \bigoplus_{\chi\in\mathcal{E}} \operatorname{Skew}(2,\mathbb{C}) \Big\}$$

$$= \sum_{\chi\in\mathcal{E}} \inf\Big\{ \big\| \widehat{\nabla_{\mathcal{R}'} u}(\chi) - g_{\mathcal{R}'}(S) \big\|^2 \,\Big|\, S \in \operatorname{Skew}(2,\mathbb{C}) \Big\}$$

$$= \sum_{k\in[0,1)\cap\mathbb{Q}} \inf\Big\{ \big\| \big(\chi_k(h)^{-1}\widehat{u}(\chi_k) - L(h)^{\mathsf{T}}\widehat{u}(\chi_k) \big.$$

$$\Big. - L(h)^{\mathsf{T}} \big(\begin{smallmatrix} 0 & -s \\ s & 0 \end{smallmatrix} \big)(h \cdot x_0 - x_0) \big)_{h\in\mathcal{R}'} \big\|^2 \,\Big|\, s \in \mathbb{C} \Big\}$$

$$= \sum_{k\in[0,1)\cap\mathbb{Q}} \inf\Big\{ \sum_{n=1}^{3} \big\| e^{-2\pi i n k}\widehat{u}(\chi_k) - \big(\begin{smallmatrix} -1 & 0 \\ 0 & 1 \end{smallmatrix} \big)^n \widehat{u}(\chi_k)$$

$$- \big(\begin{smallmatrix} (-1)^{n+1} n s \\ ((-1)^n - 1) s \end{smallmatrix} \big) \big\|^2 \,\Big|\, s \in \mathbb{C} \Big\}, \tag{3.46}$$

where we used Proposition 2.56 in the second step and Lemma 2.58 in the fourth step.

It holds

$$\sum_{i=1}^{n} a_i^2 \leq \Big(\sum_{i=1}^{n} a_i \Big)^2 \leq n \sum_{i=1}^{n} a_i^2 \tag{3.47}$$

for all $n \in \mathbb{N}$ and $a_1, \ldots, a_n \geq 0$.
We define the functions

$$f_1 \colon [0,1) \times \mathbb{C}^2 \times \mathbb{C} \to [0, \infty)$$

$$(k, v, s) \mapsto \sum_{n=1}^{3} \left\| e^{-2\pi i n k} v - \begin{pmatrix} -1 & 0 \\ 0 & 1 \end{pmatrix}^n v - \begin{pmatrix} (-1)^{n+1} n s \\ ((-1)^n - 1) s \end{pmatrix} \right\|$$

and

$$f_2 \colon [0,1) \times \mathbb{C}^2 \to [0, \infty)$$

$$(k, v) \mapsto |k - \tfrac{1}{2}|_1^2 |v_1| + |k|_1 |2\pi i (k - \tfrac{1}{2}) v_1 - v_2|.$$

By (3.45), (3.46) and (3.47) it suffices so show that there exist some constant $C, c > 0$ such that for all $(k, v) \in [0,1) \times \mathbb{C}^2$ we have

$$c \inf_{s \in \mathbb{C}} f_1(k, v, s) \leq f_2(k, v) \leq C \inf_{s \in \mathbb{C}} f_1(k, v, s). \tag{3.48}$$

First we show the right inequality of (3.48). Let $c_R > 0$ be small enough, e. g. $c_R = \frac{c_T^3}{400}$. Let $(k, v, s) \in [0,1) \times \mathbb{C}^2 \times \mathbb{C}$. By (3.39) and (3.40) we have

$$\begin{aligned}
f_1(k, v, s) &\geq \left| e^{-2\pi i k} v_1 + v_1 - s \right| + \tfrac{1}{2} \left| e^{-4\pi i k} v_1 - v_1 + 2s \right| \\
&\geq \tfrac{1}{2} \left| 2(e^{-2\pi i k} v_1 + v_1 - s) + e^{-4\pi i k} v_1 - v_1 + 2s \right| \\
&= \tfrac{1}{2} \left| e^{-2\pi i k} + 1 \right|^2 |v_1| \\
&\geq \tfrac{c_T^2}{2} |k - \tfrac{1}{2}|_1^2 |v_1| \tag{3.49}
\end{aligned}$$

and

$$\begin{aligned}
f_1(k, v, s) &\geq \sum_{n \in \{1,3\}} \left| e^{-2\pi i n k} v_2 - v_2 + 2s \right| \\
&\geq \left| e^{-2\pi i k} v_2 - v_2 + 2s - (e^{-6\pi i k} v_2 - v_2 + 2s) \right| \\
&= \left| e^{-2\pi i k} + 1 \right| \left| e^{-2\pi i k} - 1 \right| |v_2| \\
&\geq c_T^2 |k|_1 |k - \tfrac{1}{2}|_1 |v_2|. \tag{3.50}
\end{aligned}$$

Case 1: $k \in [0, \tfrac{1}{4}] \cup [\tfrac{3}{4}, 1)$.
Since $k \in [0, \tfrac{1}{4}] \cup [\tfrac{3}{4}, 1)$, we have $|k - \tfrac{1}{2}|_1 \geq \tfrac{1}{4}$. By (3.49) and (3.50) we have

$$f_1(k, v, s) \geq c_R |k - \tfrac{1}{2}|_1^2 |v_1| + \pi c_R |v_1| + c_R |k|_1 |v_2| \geq c_R f_2(k, v),$$

where in the last step we used the triangle inequality.

Case 2: $k \in (\frac{1}{4}, \frac{3}{4})$.

Since $k \in (\frac{1}{4}, \frac{3}{4})$, we have $|k|_1 \geq \frac{1}{4}$. By (3.39) and (3.43) we have

$$
\begin{aligned}
f_1(k, v, s) &\geq \left|(e^{-4\pi i k} - 1)v_1 + 2s\right| + \left|(e^{-2\pi i k} - 1)v_2 + 2s\right| \\
&\geq \left|(e^{-4\pi i k} - 1)v_1 + 2s - ((e^{-2\pi i k} - 1)v_2 + 2s)\right| \\
&= \left|e^{-2\pi i k} - 1\right| \left|(e^{-2\pi i k} + 1)v_1 - v_2\right| \\
&\geq \tfrac{c_T}{4}\left|(e^{-2\pi i k} + 1)v_1 - v_2\right| \\
&\geq \tfrac{c_T}{4}\left|2\pi i(k - \tfrac{1}{2})v_1 - v_2\right| - \tfrac{c_T}{4}\left|e^{-2\pi i k} + 1 - 2\pi i(k - \tfrac{1}{2})\right||v_1| \\
&\geq \tfrac{c_T}{4}\left|2\pi i(k - \tfrac{1}{2})v_1 - v_2\right| - \tfrac{1}{4}\left|k - \tfrac{1}{2}\right|_1^2 |v_1|. \tag{3.51}
\end{aligned}
$$

By (3.49) and (3.51) we have $f_1(k, v, s) \geq c_R f_2(k, v)$.

Now we show the left inequality of (3.48). Let $C_L > 0$ be large enough, e.g. $C_L = \frac{120}{c_T}$. Let $(k, v) \in [0, 1) \times \mathbb{C}^2$. We have

$$
f_2(k, v) \geq |k|_1 \left|k - \tfrac{1}{2}\right|_1 \left|2\pi i(k - \tfrac{1}{2})v_1 - v_2\right| \geq |k|_1 \left|k - \tfrac{1}{2}\right|_1 |v_2| - \pi \left|k - \tfrac{1}{2}\right|_1^2 |v_1|. \tag{3.52}
$$

By (3.52) and the definition of f_2, we have

$$
f_2(k, v) \geq \tfrac{1}{5}|k|_1 \left|k - \tfrac{1}{2}\right|_1 |v_2|. \tag{3.53}
$$

Case 1: $k \in [0, \frac{1}{4}] \cup [\frac{3}{4}, 1)$.

Since $k \in [0, \frac{1}{4}] \cup [\frac{3}{4}, 1)$, we have $|k - \frac{1}{2}|_1 \geq \frac{1}{4}$. We have

$$
\begin{aligned}
\inf_{s \in \mathbb{C}} f_1(k, v, s) &\leq f_1(k, v, 0) \\
&\leq 6|v_1| + |v_2| \sum_{n=1}^{3} \left|e^{-2\pi i n k} - 1\right| \\
&= 6|v_1| + \left|e^{-2\pi i k} - 1\right| |v_2| \left|\sum_{n=1}^{3} \sum_{m=0}^{n-1} e^{-2\pi i m k}\right| \\
&\leq 6|v_1| + \tfrac{6}{c_T}|k|_1 |v_2| \\
&\leq C_L f_2(k, v),
\end{aligned}
$$

where we used (3.41) in the second to last step and (3.53) in the last step.

Case 2: $k \in (\frac{1}{4}, \frac{3}{4})$.

Since $k \in (\frac{1}{4}, \frac{3}{4})$, we have $|k|_1 \geq \frac{1}{4}$. By (3.43) and (3.42) we have

$$\inf_{s \in \mathbb{C}} f_1(k, v, s) \leq f_1(k, v, v_2)$$

$$\leq \sum_{n=1}^{3} \left(\left| (e^{-2\pi i n k} - (-1)^n) v_1 + (-1)^n n v_2 \right| + \left| e^{-2\pi i n k} - (-1)^n \right| |v_2| \right)$$

$$\leq \sum_{n=1}^{3} \left(\left| e^{-2\pi i n k} - (-1)^n + (-1)^n 2\pi i n (k - \tfrac{1}{2}) \right| |v_1| \right.$$

$$\left. + n \left| 2\pi i (k - \tfrac{1}{2}) v_1 - v_2 \right| + \left| e^{-2\pi i n k} - (-1)^n \right| |v_2| \right)$$

$$\leq \frac{6}{c_T} \left(\left| k - \tfrac{1}{2} \right|_1^2 |v_1| + \left| 2\pi i (k - \tfrac{1}{2}) v_1 - v_2 \right| + \left| k - \tfrac{1}{2} \right|_1 |v_2| \right). \qquad (3.54)$$

By (3.53) and (3.54) we have

$$\inf_{s \in \mathbb{C}} f_1(k, v, s) \leq C_L f_2(k, v). \qquad \square$$

4. Stability of objective structures

We use the following notation. Let d, d_1, d_2, $\mathcal{G}$, $\mathcal{T}$ and $\mathcal{F}$ be as in Definition 2.6, M_0 as in Definition 2.13 and $\mathcal{C}_N$ as in Definition 2.50 for all $N \in M_0$. We assume that the group $\mathcal{G}$ is not trivial. Let $x_0 \in \mathbb{R}^d$ be such that the map $\mathcal{G} \to \mathbb{R}^d$, $g \mapsto g \cdot x_0$ is injective. Let d_{aff} denote the dimension $\dim(\mathcal{G} \cdot x_0)$. Moreover we suppose that

$$\mathrm{aff}(\mathcal{G} \cdot x_0) = \{0_{d-d_{\mathrm{aff}}}\} \times \mathbb{R}^{d_{\mathrm{aff}}},$$

which can be achieved by a coordinate transformation, see Lemma 2.84. Let $\mathcal{R} \subset \mathcal{G}$ be such that $\mathcal{R}$ has Property 2.

For all sets $\mathcal{H} \subset \mathcal{G}$ we define a group action of $\mathrm{O}(d)$ on $\{y \colon \mathcal{H} \to \mathbb{R}^d\}$ by

$$(Ay)(g) := A(y(g)) \qquad \text{for all } A \in \mathrm{O}(d),\, y \colon \mathcal{H} \to \mathbb{R}^d \text{ and } g \in \mathcal{G}.$$

In the following we do not distinguish between the functions $\{y \colon \mathcal{G} \setminus \{id\} \to \mathbb{R}^d\}$ and the vectors $(\mathbb{R}^d)^{\mathcal{G} \setminus \{id\}}$.

4.1. The interaction potential, configurational energy and stability

Definition 4.1. Let
$$V \colon (\mathbb{R}^d)^{\mathcal{G} \setminus \{id\}} \to \mathbb{R}$$

be the *interaction potential*. We assume that V has the following properties:

(H1) (*Invariance under rotations*) For all $R \in \mathrm{SO}(d)$ and $y \colon \mathcal{G} \setminus \{id\} \to \mathbb{R}^d$ we have
$$V(Ry) = V(y).$$

(H2) (*Smoothness*) For all $y \colon \mathcal{G} \setminus \{id\} \to \mathbb{R}^d$ the function
$$L^\infty(\mathcal{G} \setminus \{id\}, \mathbb{R}^d) \to \mathbb{R}$$
$$z \mapsto V(y + z)$$

is two times continuously Fréchet differentiable, where $L^\infty(\mathcal{G} \setminus \{id\}, \mathbb{R}^d)$ is the vector space of all bounded functions from $\mathcal{G} \setminus \{id\}$ to $\mathbb{R}^d$ equipped with the uniform norm $\|\cdot\|_\infty$.

For all $y\colon \mathcal{G} \setminus \{id\} \to \mathbb{R}^d$ and $g, h \in \mathcal{G} \setminus \{id\}$ we define the partial Jacobian row vector $\partial_g V(y) \in \mathbb{R}^d$ by

$$(\partial_g V(y))_i := V'(y)(\delta_g e_i) \qquad \text{for all } i \in \{1, \dots, d\}$$

and the partial Hessian matrix $\partial_g \partial_h V(y) \in \mathbb{R}^{d \times d}$ by

$$(\partial_g \partial_h V(y))_{ij} := V''(y)(\delta_g e_i, \delta_h e_j) \qquad \text{for all } i, j \in \{1, \dots, d\},$$

where $\delta_k \colon \mathcal{G} \setminus \{id\} \to \{0, 1\}$, $l \mapsto \delta_{k,l}$ for all $k \in \mathcal{G}$.

(H3) (Summability) For all $y\colon \mathcal{G} \setminus \{id\} \to \mathbb{R}^d$ we have

$$\sum_{g \in \mathcal{G} \setminus \{id\}} \|\partial_g V(y)\| < \infty \quad \text{and} \quad \sum_{g, h \in \mathcal{G} \setminus \{id\}} \|\partial_g \partial_h V(y)\| < \infty.$$

We say a set $\mathcal{R}_V \subset \mathcal{G} \setminus \{id\}$ is an *interaction range* of V if for all for all $y\colon \mathcal{G} \setminus \{id\} \to \mathbb{R}^d$ we have $V(y) = V(\chi_{\mathcal{R}_V} y)$, where $\chi_{\mathcal{R}_V}$ is the indicator function. We say that the interaction potential V has *finite interaction range* if V has a finite interaction range. We denote $y_0 = (g \cdot x_0 - x_0)_{g \in \mathcal{G} \setminus \{id\}} \in (\mathbb{R}^d)^{\mathcal{G} \setminus \{id\}}$. If V has finite interaction range, then we extend the domain of $V'(y_0)$ and $V''(y_0)$ to $\{z \colon \mathcal{G} \setminus \{id\} \to \mathbb{R}^d\}$ and $\{z \colon \mathcal{G} \setminus \{id\} \to \mathbb{R}^d\}^2$, respectively, by

$$V'(y_0) z_1 := V'(y_0)(\chi_{\mathcal{R}_V} z_1)$$

and

$$V''(y_0)(z_1, z_2) := V''(y_0)(\chi_{\mathcal{R}_V} z_1, \chi_{\mathcal{R}_V} z_2)$$

for all $z_1, z_2 \in \{z \colon \mathcal{G} \setminus \{id\} \to \mathbb{R}^d\} \setminus L^\infty(\mathcal{G} \setminus \{id\}, \mathbb{R}^d)$, where $\mathcal{R}_V$ is a finite interaction range of V.

Remark 4.2. (i) For all functions $y\colon \mathcal{G} \setminus \{id\} \to \mathbb{R}^d$ and $z, z_1, z_2 \in L^\infty(\mathcal{G} \setminus \{id\}, \mathbb{R}^d)$ we have

$$V'(y)z = \sum_{g \in \mathcal{G} \setminus \{id\}} \partial_g V(y) z(g)$$

and

$$V''(y)(z_1, z_2) = \sum_{g, h \in \mathcal{G} \setminus \{id\}} z_1(g)^\mathsf{T} \partial_g \partial_h V(y) z_2(h).$$

(ii) In Section 4.3 and Section 4.4 we assume that V has finite interaction range.

(iii) If V has finite interaction range, then (H2) implies (H3).

(iv) For simplicity we assume that the domain of V is the whole space $(\mathbb{R}^d)^{\mathcal{G}\backslash\{id\}}$. It would be sufficient if V is defined only on $y_0 + U$, where U is a small neighbourhood of $0 \in (\mathbb{R}^d)^{\mathcal{G}\backslash\{id\}}$ with respect to the uniform norm.

Example 4.3. An example of an interaction potential consisting of pair potentials is

$$V\colon (\mathbb{R}^d)^{\mathcal{G}\backslash\{id\}} \to \mathbb{R}, \quad y \mapsto \sum_{g\in\mathcal{G}\backslash\{id\}} v(\|y(g)\|),$$

where

$$v\colon (0,\infty) \to \mathbb{R}, \quad r \mapsto r^{-12} - r^{-6}$$

is the Lennard-Jones potential.

As we have seen in Section 3.1, in our physical model we have a canonical bijection between $\mathcal{G}$ and the atoms. For a given displacement $u\colon \mathcal{G} \to \mathbb{R}^d$ the atoms are at the points $(g \cdot (x_0 + u(g))_{g\in\mathcal{G}}$ and in particular $u = 0$ is the identity. Only in the following definition and in Definition 4.8, in contrast to the remainder of this thesis, the physical model is that for a given deformation $u\colon \mathcal{G} \to \mathbb{R}^d$ the atoms are at the points $(g \cdot u(g))_{g\in\mathcal{G}}$ and in particular $u = \chi_{\mathcal{G}} x_0$ is the identity.

Definition 4.4. Let

$$E\colon U_{\mathrm{per}} \to \mathbb{R}$$
$$u \mapsto \frac{1}{|\mathcal{C}_N|} \sum_{g\in\mathcal{C}_N} V\Big(\big((gh)\cdot u(gh) - g\cdot u(g)\big)_{h\in\mathcal{G}\backslash\{id\}}\Big),$$

where u is $\mathcal{T}^N$-periodic and $N \in M_0$, be the *configurational energy*.

Remark 4.5. The function E is well-defined and independent of the choice of the representation set $\mathcal{C}_N$ for all $N \in M_0$.

Lemma 4.6. *The function E is two times continuously Fréchet differentiable with respect to the uniform norm $\|\cdot\|_\infty$. We have*

$$E(\chi_{\mathcal{G}} x_0) = V(y_0),$$
$$E'(\chi_{\mathcal{G}} x_0)u = \frac{1}{|\mathcal{C}_N|} \sum_{g\in\mathcal{C}_N} V'(y_0)\big(L(h)u(gh) - u(g)\big)_{h\in\mathcal{G}\backslash\{id\}},$$

and

$$E''(\chi_{\mathcal{G}} x_0)(u,v) = \frac{1}{|\mathcal{C}_N|} \sum_{g \in \mathcal{C}_N} V''(y_0)\Big(\big(L(h)u(gh) - u(g)\big)_{h \in \mathcal{G} \setminus \{id\}},$$

$$\big(L(h)v(gh) - v(g)\big)_{h \in \mathcal{G} \setminus \{id\}}\Big)$$

for all $u, v \in U_{\mathrm{per}}$ and $N \in M_0$ such that u and v are $\mathcal{T}^N$-periodic.

Proof. By (H1) we have

$$E(u) = \frac{1}{|\mathcal{C}_N|} \sum_{g \in \mathcal{C}_N} V\Big(\big(h \cdot u(gh) - u(g)\big)_{h \in \mathcal{G} \setminus \{id\}}\Big) \qquad (4.1)$$

for all $u \in U_{\mathrm{per}}$ and $N \in M_0$ such that u is $\mathcal{T}^N$-periodic. By (H2) the function V is two times Fréchet differentiable. We define the vector space

$$W = \Big\{ w \colon \mathcal{G} \to L^\infty(\mathcal{G} \setminus \{id\}, \mathbb{R}^d) \,\Big|\, w \text{ is periodic} \Big\}$$

and equip U_{per} and W each with the uniform norm $\|\cdot\|_\infty$. The linear map

$$\varphi_1 \colon U_{\mathrm{per}} \to W$$

$$u \mapsto \Big(\mathcal{G} \to L^\infty(\mathcal{G} \setminus \{id\}, \mathbb{R}^d), g \mapsto (L(h)u(gh) - u(g))_{h \in \mathcal{G} \setminus \{id\}} \Big)$$

is bounded and thus two times continuously Fréchet differentiable. The first and second derivative of the function

$$\varphi_2 \colon W \to \mathbb{R}$$

$$w \mapsto \frac{1}{|\mathcal{C}_N|} \sum_{g \in \mathcal{C}_N} V\big((\tau(h))_{h \in \mathcal{G} \setminus \{id\}} + w(g)\big) \quad \text{if } w \text{ is } \mathcal{T}^N\text{-periodic}$$

is given by

$$\varphi_2'(w)w_1 = \frac{1}{|\mathcal{C}_N|} \sum_{g \in \mathcal{C}_N} V'\big((\tau(h))_{h \in \mathcal{G} \setminus \{id\}} + w(g)\big)w_1(g)$$

and

$$\varphi_2''(w)(w_1, w_2) = \frac{1}{|\mathcal{C}_N|} \sum_{g \in \mathcal{C}_N} V''\big((\tau(h))_{h \in \mathcal{G} \setminus \{id\}} + w(g)\big)(w_1(g), w_2(g))$$

for all $w, w_1, w_2 \in W$ and $N \in M_0$ such that w, w_1 and w_2 are $\mathcal{T}^N$-periodic. Thus φ_2 is two times continuously Fréchet differentiable. Since

$E = \varphi_2 \circ \varphi_1$, also the function E is two times continuously Fréchet differentiable.

Equation (4.1) also implies the representations of $E(\chi_{\mathcal{G}}x_0)$, $E'(\chi_{\mathcal{G}}x_0)$ and $E''(\chi_{\mathcal{G}}x_0)$. $\qquad\square$

Remark 4.7. (i) If the map in (H2) is n times (continuously) Fréchet differentiable for some natural number n, then also E is n times (continuously) Fréchet differentiable with respect to the uniform norm $\|\cdot\|_\infty$. The proof is analogous.

(ii) The function E need not be continuous with respect to the norm $\|\cdot\|_2$. In particular E is not two times Fréchet differentiable with respect to $\|\nabla_{\mathcal{R}}\cdot\|_2$ although in other models a similar proposition is true, see, e.g., [48, Theorem 1].

Definition 4.8. We say that $u \in U_{\text{per}}$ is a *critical point* of E if $E'(u) = 0$. We say that $(\mathcal{G}, x_0, V)$ is *stable (in the atomistic model)* with respect to $\|\cdot\|_{\mathcal{R}}$ (resp. $\|\cdot\|_{\mathcal{R},0,0}$) if $\chi_{\mathcal{G}}x_0$ is a critical point of E and the bilinear form $E''(\chi_{\mathcal{G}}x_0)$ is coercive with respect to $\|\cdot\|_{\mathcal{R}}$ (resp. $\|\cdot\|_{\mathcal{R},0,0}$), i.e. there exists a constant $c > 0$ such that

$$c\|u\|_{\mathcal{R}}^2 \le E''(\chi_{\mathcal{G}}x_0)(u,u) \qquad \text{for all } u \in U_{\text{per}}.$$

We define the constants

$$\lambda_{\text{a}} := \sup\{c \in \mathbb{R} \mid \forall u \in U_{\text{per}} : c\|u\|_{\mathcal{R}}^2 \le E''(\chi_{\mathcal{G}}x_0)(u,u)\} \in \mathbb{R} \cup \{-\infty\}$$

and

$$\lambda_{\text{a},0,0} := \sup\{c \in \mathbb{R} \mid \forall u \in U_{\text{per}} : c\|u\|_{\mathcal{R},0,0}^2 \le E''(\chi_{\mathcal{G}}x_0)(u,u)\}$$
$$\in \mathbb{R} \cup \{-\infty\}.$$

Remark 4.9. (i) The bilinear form $E''(\chi_{\mathcal{G}}x_0)$ is coercive with respect to the seminorm $\|\cdot\|_{\mathcal{R}}$ (resp. $\|\cdot\|_{\mathcal{R},0,0}$) if and only if $\lambda_{\text{a}} > 0$ (resp. $\lambda_{\text{a},0,0} > 0$).

(ii) If $(\mathcal{G}, x_0, V)$ is stable with respect to $\|\cdot\|_{\mathcal{R},0,0}$, then $(\mathcal{G}, x_0, V)$ is also stable with respect to $\|\cdot\|_{\mathcal{R}}$, see Remark 3.36.

(iii) The above definition of the stability and the constant λ_{a} generalizes the definition in [40, p. 89] where these terms are defined for lattices. For lattices we have $\lambda_{\text{a}} = \lambda_{\text{a},0,0}$ since then $\|\cdot\|_{\mathcal{R}} = \|\cdot\|_{\mathcal{R},0,0}$.

(iv) By Theorem 3.34 the stability of $(\mathcal{G}, x_0, V)$ is independent of the choice of $\mathcal{R}$.

(v) The constants λ_a and $\lambda_{a,0,0}$ need not be finite, see Example 4.40 and Example 4.41. In Section 4.4 we present sufficient conditions for both $\lambda_a \in \mathbb{R}$ and $\lambda_{a,0,0} \in \mathbb{R}$.

The following proposition states a characterization of λ_a and $\lambda_{a,0,0}$ by means of the dual problem.

Proposition 4.10. *We have*

$$\lambda_a = \inf\{E''(\chi_\mathcal{G} x_0)(u, u) \,|\, u \in U_{\mathrm{per}}, \|u\|_\mathcal{R} = 1\}$$

and

$$\lambda_{a,0,0} = \inf\{E''(\chi_\mathcal{G} x_0)(u, u) \,|\, u \in U_{\mathrm{per}}, \|u\|_{\mathcal{R},0,0} = 1\}.$$

Proof. We denote RHS $= \inf\{E''(\chi_\mathcal{G} x_0)(u, u) \,|\, u \in U_{\mathrm{per}}, \|u\|_\mathcal{R} = 1\}$. It is clear that $\lambda_a \leq$ RHS. Let $c \in \mathbb{R}$ be such that $c > \lambda_a$. There exits some $u \in U_{\mathrm{per}}$ such that $c\|u\|_\mathcal{R}^2 > E''(\chi_\mathcal{G} x_0)(u, u)$. By Theorem 3.34, Proposition 3.28 and since the group $\mathcal{G}$ is not trivial, we have $\ker(\|\cdot\|_\mathcal{R}) \neq U_{\mathrm{per}}$. Thus and since $\|\cdot\|_\mathcal{R} \leq \|\cdot\|_\infty$, we may assume that $\|u\|_\mathcal{R} = 1$. Thus we have RHS $\leq c$. Since c was arbitrary, we have $\lambda_a \geq$ RHS. The proof of the characterization of $\lambda_{a,0,0}$ is analogous. $\square$

4.2. Characterization of a critical point

Definition 4.11. We define the row vector

$$e_V := \sum_{g \in \mathcal{G}\setminus\{id\}} \partial_g V(y_0)(L(g) - I_d) \in \mathbb{R}^d$$

and the function $f_V \in L^1(\mathcal{G}, \mathbb{R}^{d \times d})$ by

$$\begin{aligned}
f_V(g) := \sum_{h_1, h_2 \in \mathcal{G}\setminus\{id\}} &\Big(\delta_{g, h_2^{-1} h_1} L(h_2)^\mathsf{T} \partial_{h_2} \partial_{h_1} V(y_0) L(h_1) \\
&- \delta_{g, h_2^{-1}} L(h_2)^\mathsf{T} \partial_{h_2} \partial_{h_1} V(y_0) - \delta_{g, h_1} \partial_{h_2} \partial_{h_1} V(y_0) L(h_1) \\
&+ \delta_{g, id} \partial_{h_2} \partial_{h_1} V(y_0)\Big)
\end{aligned}$$

for all $g \in \mathcal{G}$.

Remark 4.12.   (i) By (H3) the function f_V is well-defined and we have

$$\sum_{g \in \mathcal{G}} f_V(g) = \sum_{h_1, h_2 \in \mathcal{G}\setminus\{id\}} (L(h_2) - I_d)^\mathsf{T} \partial_{h_2} \partial_{h_1} V(y_0)(L(h_1) - I_d).$$

(ii) If $\mathcal{R}_V$ is an interaction range of V, then we have

$$\operatorname{supp} f_V \subset \mathcal{R}_V^{-1}\mathcal{R}_V \cup \mathcal{R}_V^{-1} \cup \mathcal{R}_V.$$

In particular, if V has finite interaction range, then the support of f_V is finite.

Definition 4.13. For all $N \in M_0$ and $g, h \in \mathcal{G}_N$ we define the partial Jacobian row vector $\partial_g E(\chi_{\mathcal{G}} x_0) \in \mathbb{R}^d$ by

$$(\partial_g E(\chi_{\mathcal{G}} x_0))_i := E'(\chi_{\mathcal{G}} x_0)(\delta_g e_i) \qquad \text{for all } i \in \{1, \ldots, d\}$$

and the partial Hessian matrix $\partial_g \partial_h E(\chi_{\mathcal{G}} x_0) \in \mathbb{R}^{d \times d}$ by

$$(\partial_g \partial_h E(\chi_{\mathcal{G}} x_0))_{ij} := E''(\chi_{\mathcal{G}} x_0)(\delta_g e_i, \delta_h e_j) \quad \text{for all } i, j \in \{1, \ldots, d\},$$

where $\delta_k \colon \mathcal{G} \to \{0, 1\}$, $l \mapsto \delta_{k, l \mathcal{T}^N}$ for all $k \in \mathcal{G}_N$.

The following lemma characterizes the first and second derivative of E.

Lemma 4.14. *Let $N \in M_0$. We have*

$$\partial_g E(\chi_{\mathcal{G}} x_0) = \frac{1}{|\mathcal{C}_N|} e_V \qquad \text{for all } g \in \mathcal{G}_N$$

and

$$\partial_{g_2} \partial_{g_1} E(\chi_{\mathcal{G}} x_0) = \frac{1}{|\mathcal{C}_N|} \sum_{g \in g_2^{-1} g_1} f_V(g) \qquad \text{for all } g_1, g_2 \in \mathcal{G}_N.$$

In particular we have

$$\partial_g E(\chi_{\mathcal{G}} x_0) = \partial_{id} E(\chi_{\mathcal{G}} x_0) \qquad \text{for all } g \in \mathcal{G}_N$$

and

$$\partial_{g_2} \partial_{g_1} E(\chi_{\mathcal{G}} x_0) = \partial_{id} \partial_{g_2^{-1} g_1} E(\chi_{\mathcal{G}} x_0) \qquad \text{for all } g_1, g_2 \in \mathcal{G}_N.$$

Proof. Let $N \in M_0$, $g_1, g_2 \in \mathcal{G}$ and for all $g \in \mathcal{G}_N$ let δ_g be as in Definition 4.13. Since $\mathcal{T}^N$ is a normal subgroup of $\mathcal{G}$, we have

$$\sum_{g \in \mathcal{C}_N} \delta_{g_1 \mathcal{T}^N}(gh) = \sum_{g \in \mathcal{C}_N} \sum_{t \in \mathcal{T}^N} \delta_{g_1 h^{-1}, gt} = 1 \qquad \text{for all } h \in \mathcal{G}. \quad (4.2)$$

Using Lemma 4.6, Remark 4.2(i) and (4.2), we have

$$\partial_{g_1 T^N} E(\chi_{\mathcal{G}} x_0) = \big(E'(\chi_{\mathcal{G}} x_0)(\delta_{g_1 T^N} e_i) \big)_{i \in \{1,\dots,d\}}$$

$$= \frac{1}{|\mathcal{C}_N|} \sum_{g \in \mathcal{C}_N} \sum_{h \in \mathcal{G} \setminus \{id\}} \partial_h V(y_0) \big(\delta_{g_1 T^N}(gh) L(h) - \delta_{g_1 T^N}(g) I_d \big)$$

$$= \frac{1}{|\mathcal{C}_N|} \sum_{h \in \mathcal{G} \setminus \{id\}} \partial_h V(y_0)(L(h) - I_d)$$

$$= \frac{1}{|\mathcal{C}_N|} e_V. \tag{4.3}$$

The right hand side of (4.3) is independent of $g_1 T^N$ and in particular, we have

$$\partial_{g_1 T^N} E(\chi_{\mathcal{G}} x_0) = \partial_{T^N} E(\chi_{\mathcal{G}} x_0).$$

Since T^N is a normal subgroup of $\mathcal{G}$, for all $h_1, h_2 \in \mathcal{G}$ we have

$$\sum_{g \in \mathcal{C}_N} \delta_{g_2 T^N}(gh_2) \delta_{g_1 T^N}(gh_1) = \sum_{g \in \mathcal{C}_N} \sum_{t,s \in T^N} \delta_{g_2 h_2^{-1}, gs} \delta_{g_1 h_1^{-1} t, gs}$$

$$= \sum_{t \in T^N} \delta_{g_2 h_2^{-1}, g_1 h_1^{-1} t}$$

$$= \sum_{t \in T^N} \delta_{h_2^{-1} h_1, g_2^{-1} g_1 t}. \tag{4.4}$$

Using Lemma 4.6, Remark 4.2(i) and (4.4), we have

$$\partial_{g_2 T^N} \partial_{g_1 T^N} E(\chi_{\mathcal{G}} x_0) = \big(E''(\chi_{\mathcal{G}} x_0)(\delta_{g_2 T^N} e_i, \delta_{g_1 T^N} e_j) \big)_{i,j \in \{1,\dots,d\}}$$

$$= \frac{1}{|\mathcal{C}_N|} \sum_{g \in \mathcal{C}_N} \sum_{h_1, h_2 \in \mathcal{G} \setminus \{id\}} \big(\delta_{g_2 T^N}(gh_2) L(h_2) - \delta_{g_2 T^N}(g) I_d \big)^{\mathsf{T}}$$

$$\partial_{h_2} \partial_{h_1} V(y_0) \big(\delta_{g_1 T^N}(gh_1) L(h_1) - \delta_{g_1 T^N}(g) I_d \big)$$

$$= \frac{1}{|\mathcal{C}_N|} \sum_{t \in T^N} \sum_{h_1, h_2 \in \mathcal{G} \setminus \{id\}} \Big(\delta_{h_2^{-1} h_1, g_2^{-1} g_1 t} L(h_2)^{\mathsf{T}} \partial_{h_2} \partial_{h_1} V(y_0) L(h_1)$$

$$- \delta_{h_2^{-1}, g_2^{-1} g_1 t} L(h_2)^{\mathsf{T}} \partial_{h_2} \partial_{h_1} V(y_0) - \delta_{h_1, g_2^{-1} g_1 t} \partial_{h_2} \partial_{h_1} V(y_0) L(h_1)$$

$$+ \delta_{id, g_2^{-1} g_1 t} \partial_{h_2} \partial_{h_1} V(y_0) \Big).$$

$$= \frac{1}{|\mathcal{C}_N|} \sum_{t \in T^N} f_V(g_2^{-1} g_1 t). \tag{4.5}$$

The right hand side of (4.5) is only dependent on $g_2^{-1} g_1 T^N$ and in particular, we have

$$\partial_{g_2 T^N} \partial_{g_1 T^N} E(\chi_{\mathcal{G}} x_0) = \partial_{id} \partial_{g_2^{-1} g_1 T^N} E(\chi_{\mathcal{G}} x_0). \qquad \square$$

Remark 4.15.   (i) The configurational energy is left-translation-invariant, i. e. for all $u \in U_{per}$ and $g \in \mathcal{G}$ it holds $E(u) = E(u(g \cdot))$. This implies that also $E'(\chi_{\mathcal{G}} x_0)$ and $E''(\chi_{\mathcal{G}} x_0)$ are left-translation-invariant, i. e. $E'(\chi_{\mathcal{G}} x_0) u = E'(\chi_{\mathcal{G}} x_0) u(g \cdot)$ and $E''(\chi_{\mathcal{G}} x_0)(u, v) = E(\chi_{\mathcal{G}} x_0)(u(g \cdot), v(g \cdot))$ for all $u \in U_{per}$ and $g \in \mathcal{G}$. Thus we have $\partial_{g_1} E(\chi_{\mathcal{G}} x_0) = \partial_{id} E(\chi_{\mathcal{G}} x_0)$ and $\partial_{g_2} \partial_{g_1} E(\chi_{\mathcal{G}} x_0) = \partial_{id} \partial_{g_2^{-1} g_1} E(\chi_{\mathcal{G}} x_0)$ for all $N \in M_0$ and $g_1, g_2 \in \mathcal{G}_N$.

(ii) By the above lemma we have

$$e_V = \big(E'(\chi_{\mathcal{G}} x_0)(\chi_{\mathcal{G}} e_i) \big) \big)_{i \in \{1, \dots, d\}}.$$

Now we suppose that V has finite interaction range $\mathcal{R}_V \subset \mathcal{G} \setminus \{id\}$. By Remark 4.12(ii) we have

$$\operatorname{supp} f_V \subset \mathcal{R}_V^{-1} \mathcal{R}_V \cup \mathcal{R}_V \cup \mathcal{R}_V^{-1} =: \mathcal{R}_{f_V}$$

and by the above lemma we have

$$f_V(g) = \begin{cases} |\mathcal{C}_N| \partial_{id} \partial_{g T^N} E(\chi_{\mathcal{G}} x_0) & \text{for all } g \in \mathcal{R}_{f_V} \\ 0_{d,d} & \text{else} \end{cases}$$

for all $N \in M_0$ large enough, precisely for all $N \in M_0$ such that

$$T^N \cap \mathcal{R}_{f_V}^{-1} \mathcal{R}_{f_V} \subset \{id\}.$$

Corollary 4.16. *It holds $E'(\chi_{\mathcal{G}} x_0) = 0$ if and only if $e_V = 0$*

Proof. This is clear by Lemma 4.14. $\square$

Corollary 4.17. *Suppose that $\mathcal{G} < \operatorname{Trans}(d)$. Then we have $E'(\chi_{\mathcal{G}} x_0) = 0$.*

Proof. This is clear by Corollary 4.16. $\square$

Theorem 4.18. *The triple $(\mathcal{G}, x_0, V)$ is stable with respect to $\| \cdot \|_{\mathcal{R}}$ (resp. $\| \cdot \|_{\mathcal{R},0,0}$) if and only if $e_V = 0$ and $\lambda_a > 0$ (resp. $\lambda_{a,0,0} > 0$).*

Proof. This is clear by Corollary 4.16 and Remark 4.9(i). $\square$

4.3. A sufficient condition for a minimum

As motivated in Section 3.1, the following proposition shows that $U_{\mathrm{iso}}(\mathcal{R})$ is really a tangent space.

Proposition 4.19. *There exists an open neighborhood $U \subset \mathrm{E}(d)$ of id such that the set*

$$\big\{ u\colon \mathcal{R} \to \mathbb{R}^d \,\big|\, \exists a \in U \, \forall g \in \mathcal{R} : g \cdot (x_0 + u(g)) = a \cdot (g \cdot x_0) \big\}$$

is a manifold and $U_{\mathrm{iso}}(\mathcal{R})$ is its tangent space at the point 0.

Proof. Let $B = \{S \in \mathrm{Skew}(d) \,|\, \|S\| < \log(2)\}$. By [10, Theorem 11.5.2 and Proposition 11.6.7], the matrix exponential $\exp\colon B \to \exp(B)$ is a homeomorphism and we have $\exp(B) \subset \mathrm{SO}(d)$. Let $\log$ be its inverse map. Let $U \subset \mathrm{Skew}((d - d_{\mathrm{aff}}) + d_{\mathrm{aff}})$ be a neighborhood of 0 such that the map

$$f\colon U \to \mathrm{Skew}(d)$$
$$\begin{pmatrix} S_1 & A \\ -A^{\mathsf{T}} & S_2 \end{pmatrix} \mapsto \log\!\Big(\exp\!\begin{pmatrix} 0 & A \\ -A^{\mathsf{T}} & S_2 \end{pmatrix} \exp\!\begin{pmatrix} S_1 & 0 \\ 0 & 0 \end{pmatrix} \Big)$$

is well-defined. By the inverse function theorem there exists an open neighborhood $V \subset U$ of 0 such that $W := f(V)$ is an open neighborhood of 0 and the map $f|_V\colon V \to W$ is a diffeomorphism. Without loss of generality we may assume that

$$V = \Big\{ \begin{pmatrix} S_1 & A \\ -A^{\mathsf{T}} & S_2 \end{pmatrix} \,\Big|\, S_1 \in V_1, (A, S_2) \in V_2 \Big\},$$

where $V_1 \subset \mathrm{Skew}(d - d_{\mathrm{aff}})$ is an open neighborhood of 0 and $V_2 \subset \mathbb{R}^{(d - d_{\mathrm{aff}}) \times d_{\mathrm{aff}}} \times \mathrm{Skew}(d_{\mathrm{aff}})$ is an open neighborhood of 0. The set $X := \{(\exp(A), b) \,|\, A \in W, b \in \mathbb{R}^d\} \subset \mathrm{E}(d)$ is an open neighborhood of id. We have

$$M := \big\{ u\colon \mathcal{R} \to \mathbb{R}^d \,\big|\, \exists a \in X \, \forall g \in \mathcal{R} : g \cdot (x_0 + u(g)) = a \cdot (g \cdot x_0) \big\}$$
$$= \big\{ (L(g)^{\mathsf{T}}(b + (\exp(A) - I_d)(g \cdot x_0)))_{g \in \mathcal{R}} \,\big|\, b \in \mathbb{R}^d, A \in W \big\}$$
$$= \Big\{ \Big(L(g)^{\mathsf{T}}\Big(b + \Big(\exp\!\begin{pmatrix} 0 & A \\ -A^{\mathsf{T}} & S \end{pmatrix} - I_d \Big)(g \cdot x_0 - x_0) \Big) \Big)_{g \in \mathcal{R}} \,\Big|\, b \in \mathbb{R}^d,$$
$$(A, S) \in V_2 \Big\}$$

since $g \cdot x_0 - x_0 \in \{0_{d-d_{\text{aff}}}\} \times \mathbb{R}^{d_{\text{aff}}}$ for all $g \in \mathcal{R}$. Thus the map

$$h \colon \mathbb{R}^d \times V_2 \to M$$

$$(b, A, S) \mapsto \left(L(g)^{\mathsf{T}} \left(b + \left(\exp\left(\begin{smallmatrix} 0 & A \\ -A^{\mathsf{T}} & S \end{smallmatrix} \right) - I_d \right) (g \cdot x_0 - x_0) \right) \right)_{g \in \mathcal{R}}$$

is surjective. By Lemma 3.6 there exists some $C = (c_g)_{g \in \mathcal{R}} \in \mathbb{R}^{d_{\text{aff}} \times |\mathcal{R}|}$ of rank d_{aff} such that $(g \cdot x_0 - x_0)_{g \in \mathcal{R}} = \left(\begin{smallmatrix} 0 \\ C \end{smallmatrix} \right)$. We have

$$h'(0) \colon \mathbb{R}^d \times V_2 \to (\mathbb{R}^d)^{\mathcal{R}}$$

$$(b, A, S) \mapsto \left(L(g)^{\mathsf{T}} \left(b + \left(\begin{smallmatrix} A c_g \\ S c_g \end{smallmatrix} \right) \right) \right)_{g \in \mathcal{R}}.$$

Since $id \in \mathcal{R}$ and the rank of C is equal to the number of its rows, the map $h'(0)$ is injective. Thus there exist an open neighborhood $Y \subset \mathbb{R}^d \times V_2$ of 0 and an open neighborhood $Z \subset M$ of 0 such that $h|_Y \colon Y \to Z$ is a homeomorphism. In particular M is a manifold and $U_{\text{iso}}(\mathcal{R})$ is its tangent space at 0. $\square$

Remark 4.20. A chart of the manifold of the above theorem is given in the proof.

The following theorem gives a sufficient condition for $\chi_{\mathcal{G}} x_0$ to be a minimum point of E for the cases that $\mathcal{G}$ is finite, $\mathcal{G}$ is a space group, and $d_1 = 1$.

Theorem 4.21. *Suppose that* $d_1 \in \{0, 1, d\}$, *V has finite interaction range, $e_V = 0$ and $\lambda_{a,0,0} > 0$. Then E has a local minimum point at $\chi_{\mathcal{G}} x_0$ with respect to $\| \cdot \|_\infty$, i. e. there exists a neighborhood $U \subset U_{\text{per}}$ of 0 with respect to $\| \cdot \|_\infty$ such that*

$$E(\chi_{\mathcal{G}} x_0 + u) \geq E(\chi_{\mathcal{G}} x_0) \qquad \text{for all } u \in U.$$

Proof. Suppose V has finite interaction range, $e_V = 0$ and $\lambda_{a,0,0} > 0$. First we assume that $d_1 \in \{0, 1\}$. Let $\mathcal{R}_V \subset \mathcal{G} \setminus \{id\}$ be a finite interaction range of V. Since $e_V = 0$, by Corollary 4.16 we have $E'(\chi_{\mathcal{G}} x_0) = 0$. By Theorem 3.37 there exists a constant c_1 such that $\| \cdot \|_{\mathcal{R},0,0} \geq c_1 \| \cdot \|_{\mathcal{R} \cup \mathcal{R}_V, \nabla, 0, 0}$. Let $c_2 = c_1^2 \lambda_{a,0,0} / 2 > 0$. We have

$$E''(\chi_{\mathcal{G}} x_0)(u, u) \geq \lambda_{a,0,0} \|u\|_{\mathcal{R},0,0}^2$$

$$\geq \tfrac{\lambda_{a,0,0}}{2} \|u\|_{\mathcal{R},0,0}^2 + c_2 \|u\|_{\mathcal{R} \cup \mathcal{R}_V, \nabla, 0, 0}^2$$

$$\geq \tfrac{\lambda_{a,0,0}}{2} \|u\|_{\mathcal{R},0,0}^2 + c_2 \|u\|_{\mathcal{R}_V, \nabla, 0, 0}^2$$

$$= \tfrac{\lambda_{a,0,0}}{2} \|u\|_{\mathcal{R},0,0}^2 + c_2 \|\nabla_{\mathcal{R}_V} u\|_2^2 \qquad (4.6)$$

for all $u \in U_{\text{per}}$. In the last step we used that $\| \cdot \|_{\mathcal{R}_V, \nabla, 0, 0} = \|\nabla_{\mathcal{R}_V} \cdot \|_2$ since $d_1 \in \{0, 1\}$ and thus $U_{\text{rot}, 0, 0}(\mathcal{R}_V) = \{0\}$. Since $\mathcal{R}_V$ is a finite interaction range of V, by Taylor's theorem there exists some $\varepsilon > 0$ such that for all $z \colon \mathcal{G} \setminus \{id\} \to \mathbb{R}^d$ with $\|z\|_\infty < \varepsilon$ we have

$$V(y_0 + z) \geq V(y_0) + V'(y_0)z + V''(y_0)(z, z) - c_2 \|z|_{\mathcal{R}_V}\|^2. \qquad (4.7)$$

For all $u \in U_{\text{per}}$ with $\|u\|_\infty < \varepsilon/2$ we have

$$E(\chi_{\mathcal{G}} x_0 + u) = \frac{1}{|\mathcal{C}_N|} \sum_{g \in \mathcal{C}_N} V\left(\left(h \cdot (x_0 + u(gh)) - (x_0 + u(g)) \right)_{h \in \mathcal{G} \setminus \{id\}} \right)$$

$$\geq \frac{1}{|\mathcal{C}_N|} \sum_{g \in \mathcal{C}_N} \left(V(y_0) + V'(y_0)(L(h)u(gh) - u(g))_{h \in \mathcal{G} \setminus \{id\}} \right.$$

$$+ V''(y_0)\left((L(h)u(gh) - u(g))_{h \in \mathcal{G} \setminus \{id\}}, (L(h)u(gh) - u(g))_{h \in \mathcal{G} \setminus \{id\}} \right)$$

$$\left. - c_2 \|\nabla_{\mathcal{R}_V} u(g)\|^2 \right)$$

$$= E(\chi_{\mathcal{G}} x_0) + E'(\chi_{\mathcal{G}} x_0)u + E''(\chi_{\mathcal{G}} x_0)(u, u) - c_2 \|\nabla_{\mathcal{R}_V} u\|_2^2$$

$$\geq E(\chi_{\mathcal{G}} x_0) + \frac{\lambda_{a, 0, 0}}{2} \|u\|_{\mathcal{R}, 0, 0}^2,$$

where $N \in M_0$ such that u is $\mathcal{T}^N$-periodic and we used (H1) in the first, (4.7) in the second and (4.6) in the last step.

Now we assume that $d_1 = d$, i.e. $\mathcal{G}$ is finite. Thus we have $U_{\text{iso}}(\mathcal{R}) = U_{\text{iso}, 0, 0}(\mathcal{R})$. By Proposition 4.19 there exists a neighborhood $U \subset \mathrm{E}(d)$ of id such that the set

$$M := \left\{ u \in U_{\text{per}} \,\middle|\, \exists a \in U \,\forall g \in \mathcal{G} : g \cdot (x_0 + u(g)) = a \cdot (g \cdot x_0) \right\}$$

is a manifold and $U_{\text{iso}, 0, 0}$ is its tangent space at 0. For all $u \in M$ and $v \in U_{\text{per}}$ we have

$$E(\chi_{\mathcal{G}} x_0 + u + v) = E\left(\chi_{\mathcal{G}} x_0 + \left(L(g)^{\mathsf{T}}(b + (A - I_d)(g \cdot x_0)) \right)_{g \in \mathcal{G}} + v \right)$$

$$= \frac{1}{|\mathcal{G}|} \sum_{g \in \mathcal{G}} V\left(\left((gh) \cdot \left(x_0 + L(gh)^{\mathsf{T}}(b + (A - I_d)((gh) \cdot x_0)) + v(gh) \right) \right. \right.$$

$$\left. \left. - g \cdot \left(x_0 + L(g)^{\mathsf{T}}(b + (A - I_d)(g \cdot x_0)) + v(g) \right) \right)_{h \in \mathcal{G} \setminus \{id\}} \right)$$

$$= \frac{1}{|\mathcal{G}|} \sum_{g \in \mathcal{G}} V\left(\left(A\left((gh) \cdot (x_0 + w(gh)) - g \cdot (x_0 + w(g)) \right) \right)_{h \in \mathcal{G} \setminus \{id\}} \right)$$

$$= \frac{1}{|\mathcal{G}|} \sum_{g \in \mathcal{G}} V\Bigg(\Big((gh) \cdot (x_0 + w(gh)) - g \cdot (x_0 + w(g)) \Big)_{h \in \mathcal{G} \setminus \{id\}} \Bigg)$$
$$= E(\chi_{\mathcal{G}} x_0 + w), \tag{4.8}$$

where $(A, b) \in U$ such that $g \cdot (x_0 + u(g)) = a \cdot (g \cdot x_0)$ for all $g \in \mathcal{G}$, the function $w \colon \mathcal{G} \to \mathbb{R}^d$ is defined by $g \mapsto L(g)^{\mathsf{T}} A^{\mathsf{T}} L(g) v(g)$, and we used (H1) in the second to last step. In particular we have

$$E(\chi_{\mathcal{G}} x_0 + u) = E(\chi_{\mathcal{G}} x_0) \qquad \text{for all } u \in M. \tag{4.9}$$

Since $e_V = 0$, by (4.8) and Corollary 4.16 for all $u \in M$ and $v \in U_{\text{per}}$ we have

$$
\begin{aligned}
E'(\chi_{\mathcal{G}} x_0 + u)v &= \lim_{t \to 0} \frac{E(\chi_{\mathcal{G}} x_0 + u + tv) - E(\chi_{\mathcal{G}} x_0 + u)}{t} \\
&= \lim_{t \to 0} \frac{E(\chi_{\mathcal{G}} x_0 + tw) - E(\chi_{\mathcal{G}} x_0)}{t} \\
&= E'(\chi_{\mathcal{G}} x_0)w \\
&= 0, \tag{4.10}
\end{aligned}
$$

where w is defined as above. By (4.10) we have

$$E'(\chi_{\mathcal{G}} x_0 + u) = 0 \qquad \text{for all } u \in M. \tag{4.11}$$

In the following, $c > 0$ denotes a sufficiently small constant, which may vary from line to line. Since $\lambda_{a,0,0} > 0$, we have

$$E''(\chi_{\mathcal{G}} x_0)(u, u) \geq c\|u\|_{\mathcal{R},0,0}^2 \qquad \text{for all } u \in U_{\text{per}}.$$

Let $U_{\text{iso},0,0}^{\perp}$ be the orthogonal complement of $U_{\text{iso},0,0}$ with respect to $\| \cdot \|_2$. By Theorem 3.34 the seminorm $\| \cdot \|_{\mathcal{R}}|_{U_{\text{iso},0,0}^{\perp}}$ is a norm and thus we have

$$E''(\chi_{\mathcal{G}} x_0)(u, u) \geq c\|u\|_{\infty}^2 \qquad \text{for all } u \in U_{\text{iso},0,0}^{\perp}.$$

Since E'' is continuous in $(U_{\text{per}}, \| \cdot \|_{\infty})$, without loss of generality we may assume that M is such that

$$E''(\chi_{\mathcal{G}} x_0 + u)(v, v) \geq c\|v\|_{\infty}^2 \qquad \text{for all } u \in M \text{ and } v \in U_{\text{iso},0,0}^{\perp}. \tag{4.12}$$

Without loss of generality let M be such that by (4.11), (4.12), Taylor's theorem and (4.9) there exists a neighborhood $V \subset U_{\text{iso},0,0}^{\perp}$ of 0 such that

$$E(\chi_{\mathcal{G}} x_0 + u + v) \geq E(\chi_{\mathcal{G}} x_0 + u) = E(\chi_{\mathcal{G}} x_0) \qquad \text{for all } u \in M \text{ and } v \in V.$$

Since $M + V \subset U_{\text{per}}$ is a neighborhood of 0, the assertion is proven. $\qquad \square$

Remark 4.22. Suppose that $d_1 \in \{0,1\}$, V has finite interaction range, $e_V = 0$ and $\lambda_{a,0,0} > 0$. Then there even exists a neighborhood $U \subset U_{\mathrm{per}}$ of $\chi_{\mathcal{G}} x_0$ with respect to $\| \cdot \|_\infty$ such that

$$E(\chi_{\mathcal{G}} x_0 + u) \geq E(\chi_{\mathcal{G}} x_0) + \tfrac{\lambda_{a,0,0}}{2} \|u\|_{\mathcal{R},0,0}^2 \qquad \text{for all } u \in U.$$

The above proof also shows this assertion.

4.4. Boundedness of the bilinear form $E''(\chi_{\mathcal{G}} x_0)$

In this section we present sufficient conditions for the boundedness of $E''(\chi_{\mathcal{G}} x_0)$. The boundedness of $E''(\chi_{\mathcal{G}} x_0)$ with respect to $\| \cdot \|_{\mathcal{R}}$ and $\| \cdot \|_{\mathcal{R},0,0}$ particularly implies the finiteness of λ_a and $\lambda_{a,0,0}$, respectively. With respect to $\| \cdot \|_{\mathcal{R}}$, the main result is Theorem 4.28. With respect to $\| \cdot \|_{\mathcal{R},0,0}$ and for the physical important case $d = 3$, the main results are Theorem 4.34 and Theorem 4.39. In this section we assume that V has finite interaction range.

4.4.1. The general case

We recall the definition of the boundedness of a bilinear form.

Definition 4.23. Let W be a real vector space, $\| \cdot \|$ be a seminorm on W and B be a bilinear form on W. We say the bilinear form B is *bounded* with respect to $\| \cdot \|$ if there exists a constant $C > 0$ such that

$$|B(v,w)| \leq C \|v\| \|w\| \qquad \text{for all } v, w \in U_{\mathrm{per}}.$$

The following lemma is well-known.

Lemma 4.24. *Let W be a real vector space, $\| \cdot \|$ be a seminorm on W and B be a symmetric bilinear form on W. Then B is bounded with respect to $\| \cdot \|$ if and only if there exists a constant $C > 0$ such that*

$$|B(v,v)| \leq C \|v\|^2 \qquad \text{for all } v \in W.$$

Proof. The assertion is proved in, e. g., [50, Section 92]. $\qquad\qquad\square$

Proposition 4.25. *The bilinear form $E''(\chi_{\mathcal{G}} x_0)$ is bounded with respect to $\|\nabla_{\mathcal{R}} \cdot \|_2$.*

Proof. Let $\mathcal{R}_V \subset \mathcal{G} \setminus \{id\}$ be a finite interaction range of V. By Theorem 3.40 we may assume that $\mathcal{R}_V \subset \mathcal{R}$. There exists a constant $C > 0$ such that

$$\left| V''(y_0)(z, z) \right| \leq C \| z|_{\mathcal{R}_V} \|^2 \qquad \text{for all } z \in L^\infty(\mathcal{G} \setminus \{id\}, \mathbb{R}^d). \qquad (4.13)$$

Let $u \in U_{\text{per}}$ and $N \in M_0$ such that u is $\mathcal{T}^N$-periodic. We have

$$\left| E''(\chi_{\mathcal{G}} x_0)(u, u) \right| = \left| \frac{1}{|\mathcal{C}_N|} \sum_{g \in \mathcal{C}_N} V''(y_0) \Big((L(h)u(gh) - u(g))_{h \in \mathcal{G} \setminus \{id\}}, \right.$$

$$\left. (L(h)u(gh) - u(g))_{h \in \mathcal{G} \setminus \{id\}} \Big) \right|$$

$$\leq \frac{C}{|\mathcal{C}_N|} \sum_{g \in \mathcal{C}_N} \| \nabla_{\mathcal{R}} u(g) \|^2$$

$$= C \| \nabla_{\mathcal{R}} u \|_2^2, \qquad (4.14)$$

where we used Lemma 4.6 in the first step and (4.13) in the second step. Equation (4.14) and Lemma 4.24 imply the assertion. $\qquad \square$

The property (H1) of V implies the following lemma.

Lemma 4.26. *For all* $S \in \mathrm{Skew}(d)$ *and* $z \colon \mathcal{G} \setminus \{id\} \to \mathbb{R}^d$ *we have*

$$V''(y_0)(Sy_0, z) = -V'(y_0)(Sz).$$

Proof. By (H1) for all $z \colon \mathcal{G} \setminus \{id\} \to \mathbb{R}^d$ and $A \in \mathrm{SO}(d)$ we have

$$V'(Ay_0)(Az) = \lim_{t \to 0} \frac{V(Ay_0 + tAz) - V(Ay_0)}{t} = \lim_{t \to 0} \frac{V(y_0 + tz) - V(y_0)}{t}$$

$$= V'(y_0)z. \qquad (4.15)$$

For all $S \in \mathrm{Skew}(d)$ and $z \colon \mathcal{G} \setminus \{id\} \to \mathbb{R}^d$ we have

$$V''(y_0)(Sy_0, z) = \lim_{t \to 0} \frac{V'(y_0 + tSy_0)z - V'(y_0)z}{t}$$

$$= \lim_{t \to 0} \frac{V'(e^{-tS}(y_0 + tSy_0))(e^{-tS}z) - V'(y_0)z}{t}$$

$$= \lim_{t \to 0} \frac{V'(y_0)((I_d - tS)z) - V'(y_0)z}{t}$$

$$= -V'(y_0)(Sz),$$

where we used (4.15) in the second step and Taylor's theorem in the third step. $\qquad \square$

Remark 4.27. If V does not have finite interaction range, then for all $S \in \mathrm{Skew}(d_1) \oplus \{0_{d_2,d_2}\}$ and $z \in L^\infty_{\mathrm{per}}(\mathcal{G} \setminus \{id\}, \mathbb{R}^d)$ we have

$$V''(y_0)(Sy_0, z) = -V'(y_0)(Sz).$$

The proof is analogous since we have $Sy_0 = (S(L(g)x_0 - x_0))_{g \in \mathcal{G} \setminus \{id\}} \in L^\infty(\mathcal{G} \setminus \{id\}, \mathbb{R}^d)$ for all $S \in \mathrm{Skew}(d_1) \oplus \{0_{d_2,d_2}\}$.

In the following theorem the assumption $V'(y_0) = 0$ is comparatively strong.

Theorem 4.28. *Suppose that* $V'(y_0) = 0$. *Then* $E''(\chi_{\mathcal{G}} x_0)$ *is bounded with respect to* $\| \cdot \|_{\mathcal{R}}$. *In particular we have* $\lambda_a \in \mathbb{R}$ *and* $\lambda_{a,0,0} \in \mathbb{R}$.

Proof. Suppose that $V'(y_0) = 0$. Let $u \in U_{\mathrm{per}}$ and $N \in M_0$ such that u is $\mathcal{T}^N$-periodic. Let $S \in L^\infty_{\mathrm{per}}(\mathcal{G}, \mathrm{Skew}(d))$ be $\mathcal{T}^N$-periodic such that

$$\nabla_{\mathcal{R}} u(g) = \pi_{U_{\mathrm{rot}}(\mathcal{R})}(\nabla_{\mathcal{R}} u(g)) + \left(L(h)^\mathsf{T} S(g)(h \cdot x_0 - x_0)\right)_{h \in \mathcal{R}} \text{ for all } g \in \mathcal{C}_N,$$

where $\pi_{U_{\mathrm{rot}}(\mathcal{R})}$ is the orthogonal projection on $\{v \colon \mathcal{R} \to \mathbb{R}^d\}$ with respect to the norm $\| \cdot \|$ with kernel $U_{\mathrm{rot}}(\mathcal{R})$. In the following, $C > 0$ denotes a sufficiently large constant, which is independent of u, and may vary from line to line. Let $\mathcal{R}_V \subset \mathcal{G} \setminus \{id\}$ be a finite interaction range of V. We have

$$\left|V''(y_0)(z, z)\right| \le C \|z|_{\mathcal{R}_V}\|^2 \qquad \text{for all } z \colon \mathcal{G} \setminus \{id\} \to \mathbb{R}^d. \tag{4.16}$$

We have

$$\begin{aligned}
&\left|E''(\chi_{\mathcal{G}} x_0)(u, u)\right| \\
&= \left| \frac{1}{|\mathcal{C}_N|} \sum_{g \in \mathcal{C}_N} V''(y_0)\Big(\left(L(h)u(gh) - u(g)\right)_{h \in \mathcal{G} \setminus \{id\}}, \right. \\
&\qquad \left. \left(L(h)u(gh) - u(g)\right)_{h \in \mathcal{G} \setminus \{id\}}\right) \right| \\
&= \left| \frac{1}{|\mathcal{C}_N|} \sum_{g \in \mathcal{C}_N} V''(y_0)\Big(\left(L(h)u(gh) - u(g) - S(g)(h \cdot x_0 - x_0)\right)_{h \in \mathcal{G} \setminus \{id\}}, \right. \\
&\qquad \left. \left(L(h)u(gh) - u(g) - S(g)(h \cdot x_0 - x_0)\right)_{h \in \mathcal{G} \setminus \{id\}}\right) \right| \\
&\le \frac{C}{|\mathcal{C}_N|} \sum_{g \in \mathcal{C}_N} \left\|\pi_{U_{\mathrm{rot}}(\mathcal{R})}(\nabla_{\mathcal{R}} u(g))\right\|^2
\end{aligned}$$

$$= C\|u\|_{\mathcal{R},\nabla}^2$$
$$\leq C\|u\|_{\mathcal{R}}^2, \tag{4.17}$$

where we used Lemma 4.6 in the first step, Lemma 4.26 in the second step, (4.16) in the third step and Theorem 3.34 in the last step. Lemma 4.24 and (4.17) imply the assertion. $\qquad\square$

Proposition 4.29. *Suppose that* $E'(\chi_{\mathcal{G}}x_0) = 0$. *Then we have*

$$E''(\chi_{\mathcal{G}}x_0)(u, v) = 0 \qquad \text{for all } u \in U_{\text{iso},0,0} \cap U_{\text{per}} \text{ and } v \in U_{\text{per}}.$$

Proof. Suppose that $E'(\chi_{\mathcal{G}}x_0) = 0$. Let $u \in U_{\text{iso},0,0} \cap U_{\text{per}}$ and $v \in U_{\text{per}}$. There exist some $a \in \mathbb{R}^d$ and $S \in \text{Skew}(d_1) \oplus \{0_{d_2,d_2}\}$ such that

$$L(g)u(g) = a + S(g \cdot x_0 - x_0) \qquad \text{for all } g \in \mathcal{G}.$$

Let $N \in M_0$ such that u and v are $\mathcal{T}^N$-periodic. We have

$$E''(\chi_{\mathcal{G}}x_0)(u, v)$$
$$= \frac{1}{|\mathcal{C}_N|} \sum_{g \in \mathcal{C}_N} V''(y_0)\Big(\big(L(h)u(gh) - u(g)\big)_{h \in \mathcal{G}\setminus\{id\}},$$
$$\big(L(h)v(gh) - v(g)\big)_{h \in \mathcal{G}\setminus\{id\}}\Big)$$
$$= \frac{1}{|\mathcal{C}_N|} \sum_{g \in \mathcal{C}_N} V''(y_0)\Big(\big(L(g)^\mathsf{T} SL(g)(h \cdot x_0 - x_0)\big)_{h \in \mathcal{G}\setminus\{id\}},$$
$$\big(L(h)v(gh) - v(g)\big)_{h \in \mathcal{G}\setminus\{id\}}\Big)$$
$$= -\frac{1}{|\mathcal{C}_N|} \sum_{g \in \mathcal{C}_N} V'(y_0)\big(L(g)^\mathsf{T} SL(g)(L(h)v(gh) - v(g))\big)_{h \in \mathcal{G}\setminus\{id\}}$$
$$= -\frac{1}{|\mathcal{C}_N|} V'(y_0)\Big(\big(L(h) - I_d\big) \sum_{g \in \mathcal{C}_N} L(g)^\mathsf{T} SL(g)v(g)\Big)_{h \in \mathcal{G}\setminus\{id\}}$$
$$= 0,$$

where we used Lemma 4.26 in the third step, for all $h \in \mathcal{G} \setminus \{id\}$ the equality

$$\sum_{g \in \mathcal{C}_N} L(g)^\mathsf{T} SL(g)L(h)v(gh) = \sum_{g \in \mathcal{C}_N} L(gh^{-1})^\mathsf{T} SL(gh^{-1})L(h)v(g)$$
$$= \sum_{g \in \mathcal{C}_N} L(h)L(g)^\mathsf{T} SL(g)v(g)$$

in the fourth step and Corollary 4.16 in the last step. $\qquad\square$

Remark 4.30.   (i) In the above proposition the assumption $E'(\chi_{\mathcal{G}} x_0) = 0$ is necessary, see Example 4.40.

  (ii) In the above proposition the assumption V has finite interaction range is not necessary. Using Remark 4.27 instead of Lemma 4.26, the proof is analogous. See also Lemma A.3.

 (iii) If V is weakly* sequentially continuous in addition to the above assumptions, then we also have $\frac{d^3}{d\tau^3} E(\chi_{\mathcal{G}} x_0 + \tau u)\big|_{\tau=0} = 0$ for all $u \in U_{\text{iso},0,0} \cap U_{\text{per}}$, see Proposition A.3.

4.4.2. The case $d = d_1$, i.e. $\mathcal{G}$ is finite

Theorem 4.31. *Suppose that $\mathcal{G}$ is finite and $E'(\chi_{\mathcal{G}} x_0) = 0$. Then $E''(\chi_{\mathcal{G}} x_0)$ is bounded with respect to $\|\cdot\|_{\mathcal{R}}$. In particular we have $\lambda_a = \lambda_{a,0,0} \in \mathbb{R}$.*

Proof. Suppose that $\mathcal{G}$ is finite and $E'(\chi_{\mathcal{G}} x_0) = 0$. In particular $\lambda_{\text{a}} = \lambda_{\text{a},0,0} \in \mathbb{R}$ since $\mathcal{G}$ being finite entails $\|\cdot\|_{\mathcal{R}} = \|\cdot\|_{\mathcal{R},0,0}$. Let U be a subspace of U_{per} such that $U_{\text{per}} = U_{\text{iso},0,0} \oplus U$. By Theorem 3.34 the seminorm $\|\cdot\|_{\mathcal{R}}$ is a norm on U and thus there exists a constant $C > 0$ such that $\|\cdot\|_{\infty} \leq C\|\cdot\|_{\mathcal{R}}$ on U. We have

$$\sup\{|E''(\chi_{\mathcal{G}} x_0)(u,u)| \,\big|\, u \in U_{\text{per}}, \|u\|_{\mathcal{R}} \leq 1\}$$
$$= \sup\{|E''(\chi_{\mathcal{G}} x_0)(u,u)| \,\big|\, u \in U, \|u\|_{\mathcal{R}} \leq 1\}$$
$$\leq \sup\{|E''(\chi_{\mathcal{G}} x_0)(u,u)| \,\big|\, u \in U, \|u\|_{\infty} \leq C\}$$
$$< \infty,$$

where we used Proposition 4.29 and Theorem 3.34 in the first step and in the last step that $E''(\chi_{\mathcal{G}} x_0)$ is bounded with respect to $\|\cdot\|_{\infty}$ by Lemma 4.6. With Lemma 4.24 the assertion follows.   □

4.4.3. The case $d = d_2$, i.e. $\mathcal{G}$ is a space group

If $\mathcal{G}$ is a space group, then by Corollary 3.42 there exists a constant $c > 0$ such that

$$c\lambda_{\text{a}} \leq \lambda_{\text{a},0,0} \leq \tfrac{1}{c}\lambda_{\text{a}}$$

and thus it is sufficient to consider only λ_{a}.

Theorem 4.32. *Suppose that $\mathcal{G}$ is a space group. Then $E''(\chi_{\mathcal{G}} x_0)$ is bounded with respect to both $\|\cdot\|_{\mathcal{R}}$ and $\|\cdot\|_{\mathcal{R},0,0}$. In particular we have $\lambda_a, \lambda_{a,0,0} \in \mathbb{R}$.*

Proof. This is clear by Proposition 4.25 and Corollary 3.42.   □

4.4.4. The case $d = 1 + d_2$

Theorem 4.33. *Suppose that $d = 1 + d_2$. Then $E''(\chi_{\mathcal{G}} x_0)$ is bounded with respect to $\|\cdot\|_{\mathcal{R},0,0}$. In particular we have $\lambda_{a,0,0} \in \mathbb{R}$.*

Proof. Suppose that $d = 1 + d_2$. Then we have $U_{\mathrm{rot},0,0}(\mathcal{R}) = \{0\}$ and thus $\|\cdot\|_{\mathcal{R},\nabla,0,0} = \|\nabla_{\mathcal{R}} \cdot\|_2$. With Theorem 3.37 and Proposition 4.25 follows the assertion. $\qquad\square$

4.4.5. The case $d \in \{1, 2, 3\}$

Theorem 4.34. *Suppose that $d \in \{1, 2, 3\}$, $(d, d_2) \neq (3, 1)$ and $E'(\chi_{\mathcal{G}} x_0) = 0$. Then $E''(\chi_{\mathcal{G}} x_0)$ is bounded with respect to $\|\cdot\|_{\mathcal{R},0,0}$. In particular we have $\lambda_{a,0,0} \in \mathbb{R}$.*

Proof. This is clear by Theorem 4.31, Theorem 4.32 and Theorem 4.33. $\qquad\square$

Remark 4.35. For the case $(d, d_2) = (3, 1)$ see Theorem 4.39.

4.4.6. The case $d = 2 + d_2$

For the proof of Theorem 4.39 we need the following definition.

Definition 4.36. For all $u \in U_{\mathrm{per}}$ we define the function S_u of

$$L^\infty\left(\mathcal{G}, \left\{ \begin{pmatrix} 0 & A_1 & 0 \\ -A_1^\mathsf{T} & A_2 & 0 \\ 0 & 0 & 0 \end{pmatrix} \,\middle|\, A_1 \in \mathbb{R}^{(d - d_{\mathrm{aff}}) \times (d_{\mathrm{aff}} - d_2)}, A_2 \in \mathrm{Skew}(d_{\mathrm{aff}} - d_2) \right\} \right)$$

by the condition

$$\nabla_{\mathcal{R}} u(g) = \pi_{U_{\mathrm{rot},0,0}(\mathcal{R})}(\nabla_{\mathcal{R}} u(g)) + \left(L(gh)^\mathsf{T} S_u(g) L(g)(h \cdot x_0 - x_0) \right)_{h \in \mathcal{R}}$$

for all $g \in \mathcal{G}$, where $\pi_{U_{\mathrm{rot},0,0}(\mathcal{R})}$ is the orthogonal projection on $\{v \colon \mathcal{R} \to \mathbb{R}^d\}$ with respect to the norm $\|\cdot\|$ with kernel $U_{\mathrm{rot},0,0}(\mathcal{R})$.

Remark 4.37. For all $u \in U_{\mathrm{per}}$ the function S_u is well-defined: Let $g \in \mathcal{G}$. By Lemma 2.85 there exist $B_1 \in \mathrm{O}(d - d_{\mathrm{aff}})$, $B_2 \in \mathrm{O}(d_{\mathrm{aff}} - d_2)$ and $B_3 \in \mathrm{O}(d_2)$ such that $L(g) = B_1 \oplus B_2 \oplus B_3$. By Proposition 3.28 we have

$$U_{\mathrm{rot},0,0}(\mathcal{R})$$
$$= \left\{ \mathcal{R} \to \mathbb{R}^d, h \mapsto L(h)^\mathsf{T} \left(\begin{pmatrix} 0 & A_1 \\ -A_1^\mathsf{T} & A_2 \end{pmatrix} \oplus 0_{d_2, d_2} \right)(h \cdot x_0 - x_0) \,\middle|\, \right.$$
$$\left. (A_1, A_2) \in T \right\}$$

$$= \left\{ \mathcal{R} \to \mathbb{R}^d, h \mapsto L(h)^{\mathsf{T}} \left(\begin{pmatrix} 0 & B_1^{\mathsf{T}} A_1 B_2 \\ -B_2^{\mathsf{T}} A_1^{\mathsf{T}} B_1 & B_2^{\mathsf{T}} A_2 B_2 \end{pmatrix} \oplus 0_{d_2, d_2} \right) (h \cdot x_0 - x_0) \; \middle| \right.$$
$$\left. (A_1, A_2) \in T \right\}$$
$$= \left\{ \mathcal{R} \to \mathbb{R}^d, h \mapsto L(gh)^{\mathsf{T}} \left(\begin{pmatrix} 0 & A_1 \\ -A_1^{\mathsf{T}} & A_2 \end{pmatrix} \oplus 0_{d_2, d_2} \right) L(g)(h \cdot x_0 - x_0) \; \middle| \right.$$
$$\left. (A_1, A_2) \in T \right\},$$

where $T = \mathbb{R}^{(d - d_{\mathrm{aff}}) \times (d_{\mathrm{aff}} - d_2)} \times \mathrm{Skew}(d_{\mathrm{aff}} - d_2)$.

Lemma 4.38. *For all $g_0 \in \mathcal{G}$ there exists a constant $C > 0$ such that*

$$\frac{1}{|\mathcal{C}_N|} \sum_{g \in \mathcal{C}_N} \| S_u(gg_0) - S_u(g) \|^2 \leq C \| u \|_{\mathcal{R}, 0, 0}^2$$

for all $u \in U_{\mathrm{per}}$ and $N \in M_0$ such that u is $\mathcal{T}^N$-periodic.

Proof. Let $g_0 \in \mathcal{G}$. By Lemma 3.6 there exists some $\mathcal{R}' \subset \mathcal{R}$ and $A \in \mathrm{GL}(d_{\mathrm{aff}})$ such that

$$(g \cdot x_0 - x_0)_{g \in \mathcal{R}'} = \begin{pmatrix} 0 \\ A \end{pmatrix}.$$

By Theorem 3.37 without loss of generality, we may assume that $\{g_0\} \cup g_0 \mathcal{R}' \subset \mathcal{R}$. Let $u \in U_{\mathrm{per}}$ and $N \in M_0$ such that u is $\mathcal{T}^N$-periodic. Using that $g_0 \in \mathcal{R}$ we have

$$\| u \|_{\mathcal{R}, \nabla, 0, 0}^2 = \frac{1}{|\mathcal{C}_N|} \sum_{g \in \mathcal{C}_N} \left\| \nabla_{\mathcal{R}} u(g) - \left(L(gh)^{\mathsf{T}} S_u(g) L(g)(h \cdot x_0 - x_0) \right)_{h \in \mathcal{R}} \right\|^2$$
$$\geq \frac{1}{|\mathcal{C}_N|} \sum_{g \in \mathcal{C}_N} \left\| L(g_0) u(gg_0) - u(g) - L(g)^{\mathsf{T}} S_u(g) L(g)(g_0 \cdot x_0 - x_0) \right\|^2.$$
$$(4.18)$$

Using that $g_0 \mathcal{R}' \subset \mathcal{R}$, we have

$$\| u \|_{\mathcal{R}, \nabla, 0, 0}^2 = \frac{1}{|\mathcal{C}_N|} \sum_{g \in \mathcal{C}_N} \left\| \nabla_{\mathcal{R}} u(g) - \left(L(gh)^{\mathsf{T}} S_u(g) L(g)(h \cdot x_0 - x_0) \right)_{h \in \mathcal{R}} \right\|^2$$
$$\geq \frac{1}{|\mathcal{C}_N|} \sum_{g \in \mathcal{C}_N} \sum_{h \in \mathcal{R}'} \left\| L(g_0 h) u(gg_0 h) - u(g) \right.$$
$$\left. - L(g)^{\mathsf{T}} S_u(g) L(g)((g_0 h) \cdot x_0 - x_0) \right\|^2.$$
$$(4.19)$$

Using that $\mathcal{C}_N g_0$ is a representation set of $\mathcal{G}/T^N$ and $\mathcal{R}' \subset \mathcal{R}$, we have

$$
\|u\|_{\mathcal{R},\nabla,0,0}^2 = \frac{1}{|\mathcal{C}_N|} \sum_{g \in \mathcal{C}_N} \Big\| \nabla_{\mathcal{R}} u(gg_0)
$$
$$
- \big(L(gg_0 h)^{\mathsf{T}} S_u(gg_0) L(gg_0)(h \cdot x_0 - x_0)\big)_{h \in \mathcal{R}} \Big\|^2
$$
$$
\geq \frac{1}{|\mathcal{C}_N|} \sum_{g \in \mathcal{C}_N} \sum_{h \in \mathcal{R}'} \big\| L(g_0 h) u(gg_0 h) - L(g_0) u(gg_0)
$$
$$
- L(g)^{\mathsf{T}} S_u(gg_0) L(gg_0)(h \cdot x_0 - x_0) \big\|^2. \tag{4.20}
$$

By (4.18), (4.19) and (4.20) there exists a constant $c > 0$ (independent of u and N) such that

$$
\|u\|_{\mathcal{R},\nabla,0,0}^2
$$
$$
\geq \frac{c}{|\mathcal{C}_N|} \sum_{g \in \mathcal{C}_N} \sum_{h \in \mathcal{R}'} \big\| L(g_0) u(gg_0) - u(g) - L(g)^{\mathsf{T}} S_u(g) L(g)(g_0 \cdot x_0 - x_0)
$$
$$
- L(g_0 h) u(gg_0 h) + u(g) + L(g)^{\mathsf{T}} S_u(g) L(g)((g_0 h) \cdot x_0 - x_0)
$$
$$
+ L(g_0 h) u(gg_0 h) - L(g_0) u(gg_0) - L(g)^{\mathsf{T}} S_u(gg_0) L(gg_0)(h \cdot x_0 - x_0) \big\|^2
$$
$$
= \frac{c}{|\mathcal{C}_N|} \sum_{g \in \mathcal{C}_N} \left\| (S_u(g) - S_u(gg_0)) L(gg_0) \begin{pmatrix} 0 \\ A \end{pmatrix} \right\|^2. \tag{4.21}
$$

By Lemma 2.85 for all $g \in \mathcal{C}_N$ there exist $B_1(g) \in \mathrm{O}(d - d_{\mathrm{aff}})$, $B_2(g) \in \mathrm{O}(d_{\mathrm{aff}})$, $T_1(g) \in \mathbb{R}^{(d-d_{\mathrm{aff}}) \times d_{\mathrm{aff}}}$ and $T_2(g) \in \mathrm{Skew}(d_{\mathrm{aff}})$ such that

$$
L(gg_0) = \begin{pmatrix} B_1(g) & 0 \\ 0 & B_2(g) \end{pmatrix} \quad \text{and} \quad S_u(g) - S_u(gg_0) = \begin{pmatrix} 0 & T_1(g) \\ -T_1(g) & T_2(g) \end{pmatrix}.
$$

By (4.21) we have

$$
\|u\|_{\mathcal{R},\nabla,0,0}^2 \geq \frac{c}{|\mathcal{C}_N|} \sum_{g \in \mathcal{C}_N} \Big(\|T_1(g) B_2(g) A\|^2 + \|T_2(g) B_2(g) A\|^2 \Big)
$$
$$
\geq \frac{cc_1}{|\mathcal{C}_N|} \sum_{g \in \mathcal{C}_N} \Big(\|T_1(g)\|^2 + \|T_2(g)\|^2 \Big)
$$
$$
\geq \frac{cc_1}{2|\mathcal{C}_N|} \sum_{g \in \mathcal{C}_N} \|S_u(g) - S_u(gg_0)\|^2, \tag{4.22}
$$

where $c_1 = \sigma_{\min}^2(A) > 0$, $\sigma_{\min}(M)$ denotes the minimum singular value of a matrix M and we used Corollary 9.6.7 in [10] in the second step. Theorem 3.37 and (4.22) imply the assertion. $\qquad\square$

Theorem 4.39. *Suppose that $d = 2 + d_2$, $E'(\chi_{\mathcal{G}} x_0) = 0$ and $E''(\chi_{\mathcal{G}} x_0)$ is positive semidefinite. Then $E''(\chi_{\mathcal{G}} x_0)$ is bounded with respect to $\|\cdot\|_{\mathcal{R},0,0}$.*

Proof. Suppose that $d = 2 + d_2$, $E'(\chi_{\mathcal{G}} x_0) = 0$ and $E''(\chi_{\mathcal{G}} x_0)$ is positive semidefinite. Since $E''(\chi_{\mathcal{G}} x_0)$ is positive semidefinite and by Lemma 4.24, it suffices to show that there exists a constant $C > 0$ such that

$$E''(\chi_{\mathcal{G}} x_0)(u, u) \leq C \|u\|_{\mathcal{R},0,0}^2 \qquad \text{for all } u \in U_{\text{per}}. \qquad (4.23)$$

By Theorem 2.17 there exists some $m \in \mathbb{N}$ such that $M_0 = m\mathbb{N}$. Let $\{t_1, \ldots, t_{d_2}\}$ be a generating set of $\mathcal{T}^m$. Without loss of generality we assume that

$$\mathcal{C}_{nm} = \bigcup_{n_1, \ldots, n_{d_2} \in \{0, \ldots, n-1\}} t_1^{n_1} \ldots t_{d_2}^{n_{d_2}} \mathcal{C}_m \qquad \text{for all } n \in \mathbb{N},$$

see Remark 2.51(ii). For all $g \in \mathcal{G}$ there exist $n_{1,1}, \ldots, n_{|\mathcal{C}_m|, d_2} \in \mathbb{Z}$ such that

$$\mathcal{C}_m g = \bigcup_{i=1}^{|\mathcal{C}_m|} t_1^{n_{i,1}} \ldots t_{d_2}^{n_{i,d_2}} h_i,$$

where $h_1, \ldots, h_{|\mathcal{C}_m|}$ are the elements of $\mathcal{C}_m$. Thus and since $\mathcal{T}^m$ is abelian, for all $g \in \mathcal{G}$ we have

$$\lim_{n \to \infty} \frac{|\mathcal{C}_{nm} \cap (\mathcal{C}_{nm} g)|}{|\mathcal{C}_{nm}|} = 1. \qquad (4.24)$$

Let $u \in U_{\text{per}}$ and $N \in M_0$ such that u is $\mathcal{T}^N$-periodic. For all $n \in \mathbb{N}$ we define a $\mathcal{T}^{nN}$-periodic function $v_{u,n} \in U_{\text{per}}$ by the condition

$$L(g) v_{u,n}(g) = S_u(g)(g \cdot x_0 - x_0) \qquad \text{for all } g \in \mathcal{C}_{nN}.$$

Since $\tau(\mathcal{G}) \subset \{0_{d_1}\} \times \mathbb{R}^{d_2}$, we have

$$S_u(g)(h \cdot x_0) = S_u(g) L(h) x_0 \qquad\qquad\qquad \text{for all } g, h \in \mathcal{G} \qquad (4.25)$$

and

$$S_u(g)((h_1 h_2) \cdot x_0) = S_u(g) L(h_1)(h_2 \cdot x_0) \quad \text{for all } g, h_1, h_2 \in \mathcal{G}. \qquad (4.26)$$

By (4.25) the sequence $(v_{u,n})_{n \in \mathbb{N}}$ is bounded in $(U_{\text{per}}, \|\cdot\|_\infty)$. Since the bilinear form $E''(\chi_{\mathcal{G}} x_0)$ is positive semidefinite, for all $n \in \mathbb{N}$ we have

$$E''(\chi_{\mathcal{G}} x_0)(u, u) \leq E''(\chi_{\mathcal{G}} x_0)(u, u) + E''(\chi_{\mathcal{G}} x_0)(u - 2v_{u,n}, u - 2v_{u,n})$$
$$= 2E''(\chi_{\mathcal{G}} x_0)(u - v_{u,n}, u - v_{u,n}) + 2E''(\chi_{\mathcal{G}} x_0)(v_{u,n}, v_{u,n}). \qquad (4.27)$$

In the following, $C > 0$ denotes a sufficiently large constant, which is independent of u and may vary from line to line. We have

$$\limsup_{n\to\infty} E''(\chi_\mathcal{G} x_0)(u - v_{u,n}, u - v_{u,n}) \le \limsup_{n\to\infty} C \big\|\nabla_\mathcal{R}(u - v_{u,n})\big\|_2^2$$

$$= \limsup_{n\to\infty} \frac{C}{|\mathcal{C}_{nN}|} \sum_{g\in\mathcal{C}_{nN}} \sum_{h\in\mathcal{R}} \big\|L(h)u(gh) - u(g)$$

$$- L(g)^\mathsf{T} S_u(gh)((gh)\cdot x_0 - x_0) + L(g)^\mathsf{T} S_u(g)(g\cdot x_0 - x_0)\big\|^2$$

$$\le \limsup_{n\to\infty} \frac{C}{|\mathcal{C}_{nN}|} \sum_{g\in\mathcal{C}_{nN}} \sum_{h\in\mathcal{R}} \Big(\big\|L(h)u(gh) - u(g)$$

$$- L(g)^\mathsf{T} S_u(g) L(g)(h\cdot x_0 - x_0)\big\|^2 + \big\|S_u(gh) - S_u(g)\big\|^2 \Big)$$

$$\le C\|u\|_{\mathcal{R},0,0}^2, \tag{4.28}$$

where in the first step we used Proposition 4.25, in the second step we used (4.25), (4.26) and (4.24), and in the last step we used (4.24), Lemma 4.38 and Theorem 3.37. Using (4.24) we have

$$\limsup_{n\to\infty} E''(\chi_\mathcal{G} x_0)(v_{u,n}, v_{u,n})$$

$$= \limsup_{n\to\infty} \frac{1}{|\mathcal{C}_{nN}|} \sum_{g\in\mathcal{C}_{nN}} V''(y_0)\Big((L(h)v_{u,n}(gh) - v_{u,n}(g))_{h\in\mathcal{G}\setminus\{id\}},$$

$$(L(h)v_{u,n}(gh) - v_{u,n}(g))_{h\in\mathcal{G}\setminus\{id\}}\Big)$$

$$= \limsup_{n\to\infty} \frac{1}{|\mathcal{C}_{nN}|} \sum_{g\in\mathcal{C}_{nN}} V''(y_0)\Big((a(g,h) + b(g,h))_{h\in\mathcal{G}\setminus\{id\}},$$

$$(a(g,h) + b(g,h))_{h\in\mathcal{G}\setminus\{id\}}\Big)$$

$$= \limsup_{n\to\infty}(s_{1,n} + s_{2,n}), \tag{4.29}$$

where

$$a(g,h) := L(g)^\mathsf{T}(S_u(gh) - S_u(g))((gh)\cdot x_0 - x_0),$$

$$b(g,h) := L(g)^\mathsf{T} S_u(g) L(g)(h\cdot x_0 - x_0),$$

$$s_{1,n} := \frac{1}{|\mathcal{C}_{nN}|} \sum_{g\in\mathcal{C}_{nN}} V''(y_0)\big((a(g,h))_{h\in\mathcal{G}\setminus\{id\}}, (a(g,h))_{h\in\mathcal{G}\setminus\{id\}}\big)$$

and

$$s_{2,n} := \frac{1}{|\mathcal{C}_{nN}|} \sum_{g\in\mathcal{C}_{nN}} V''(y_0)\big((b(g,h))_{h\in\mathcal{G}\setminus\{id\}}, (2a(g,h) + b(g,h))_{h\in\mathcal{G}\setminus\{id\}}\big)$$

for all $g, h \in \mathcal{G}$ and $n \in \mathbb{N}$. Let $\mathcal{R}_V \subset \mathcal{G} \setminus \{id\}$ be a finite interaction range of V. We have

$$\limsup_{n \to \infty} s_{1,n} \leq \limsup_{n \to \infty} \frac{C}{|\mathcal{C}_{nN}|} \sum_{g \in \mathcal{C}_{nN}} \sum_{h \in \mathcal{R}_V} \|a(g,h)\|^2$$

$$\leq \limsup_{n \to \infty} \frac{C}{|\mathcal{C}_{nN}|} \sum_{g \in \mathcal{C}_{nN}} \sum_{h \in \mathcal{R}_V} \|S_u(gh) - S_u(g)\|^2$$

$$\leq C\|u\|_{\mathcal{R},0,0}^2, \tag{4.30}$$

where we used (4.25) in the second and Lemma 4.38 in the last step. By Corollary 4.16, (4.24) and the boundedness of S_u, we have

$$\limsup_{n \to \infty} \frac{1}{|\mathcal{C}_{nN}|} \sum_{g \in \mathcal{C}_{nN}} V'(y_0)\big(L(g)^{\mathsf{T}} S_u(g) S_u(g) (L(g)x_0 - x_0)\big)_{h \in \mathcal{G} \setminus \{id\}}$$

$$= \limsup_{n \to \infty} \frac{1}{|\mathcal{C}_{nN}|} \sum_{g \in \mathcal{C}_{nN}} V'(y_0)\big(L(h)L(g)^{\mathsf{T}} S_u(g) S_u(g) (L(g)x_0$$

$$- x_0)\big)_{h \in \mathcal{G} \setminus \{id\}}$$

$$= \limsup_{n \to \infty} \frac{1}{|\mathcal{C}_{nN}|} \sum_{g \in \mathcal{C}_{nN}} V'(y_0)\big(L(g)^{\mathsf{T}} S_u(gh) S_u(gh) (L(gh)x_0$$

$$\tag{4.31}$$

$$- x_0)\big)_{h \in \mathcal{G} \setminus \{id\}}.$$

Since $d - d_2 = 2$, we have

$$S_u(g) S_u(h) = S_u(h) S_u(g) \qquad \text{for all } g, h \in \mathcal{G}. \tag{4.32}$$

We have

$$\limsup_{n \to \infty} s_{2,n}$$

$$= \limsup_{n \to \infty} \frac{1}{|\mathcal{C}_{nN}|} \sum_{g \in \mathcal{C}_{nN}} V'(y_0)\big(-L(g)^{\mathsf{T}} S_u(g) L(g) (2a(g,h)$$

$$+ b(g,h))\big)_{h \in \mathcal{G} \setminus \{id\}}$$

$$= \limsup_{n \to \infty} \frac{1}{|\mathcal{C}_{nN}|} \sum_{g \in \mathcal{C}_{nN}} V'(y_0)\Big(-2L(g)^{\mathsf{T}} S_u(g) S_u(gh) L(gh)x_0$$

$$+ 2L(g)^{\mathsf{T}} S_u(g) S_u(gh) x_0 + L(g)^{\mathsf{T}} S_u(g) S_u(g) L(gh)x_0$$

$$- 2L(g)^{\mathsf{T}} S_u(g) S_u(g) x_0 + L(g)^{\mathsf{T}} S_u(g) S_u(g) L(g)x_0\Big)_{h \in \mathcal{G} \setminus \{id\}}$$

$$= \limsup_{n \to \infty} \frac{1}{|\mathcal{C}_{nN}|} \sum_{g \in \mathcal{C}_{nN}} V'(y_0)\big(L(g)^{\mathsf{T}}(S_u(g) - S_u(gh))\big)^2 (L(gh)x_0$$

$$- x_0)\big)_{h \in \mathcal{G} \setminus \{id\}}$$

$$\leq \limsup_{n \to \infty} \frac{C}{|\mathcal{C}_{nN}|} \sum_{g \in \mathcal{C}_{nN}} \sum_{h \in \mathcal{R}_V} \|S_u(g) - S_u(gh)\|^2$$

$$\leq C\|u\|_{\mathcal{R},0,0}^2, \tag{4.33}$$

where in the first step we used Lemma 4.26, in the second step we used (4.25), in the third step we used (4.31) and (4.32), and in the last step we used Lemma 4.38. Equations (4.27), (4.28), (4.29), (4.30) and (4.33) imply the assertion (4.23). $\qquad\square$

4.4.7. Examples for $\lambda_a = -\infty$

In this subsection we present examples such that $\lambda_a = \lambda_{a,0,0} = -\infty$. In particular $E''(\chi_{\mathcal{G}}x_0)$ need not be bounded with respect to $\|\cdot\|_{\mathcal{R},0,0}$.

Example 4.40. We present an example such that $E''(\chi_{\mathcal{G}}x_0)(u,u) < 0$ for some $u \in U_{\text{iso},0,0} \cap U_{\text{per}}$. In particular we have $\lambda_a = \lambda_{a,0,0} = -\infty$, $E''(\chi_{\mathcal{G}}x_0)$ is not bounded with respect to $\|\cdot\|_{\mathcal{R},0,0}$, and in Proposition 4.29 and Theorem 4.31 the condition $E'(\chi_{\mathcal{G}}x_0) = 0$ is necessary. Let $d = d_2 = 2$, $p = (-I_2, 0) \in \mathrm{E}(2)$, $\mathcal{G} = \{id, p\} < \mathrm{E}(2)$, $x_0 = e_1 \in \mathbb{R}^2$ and

$$V : \mathbb{R}^2 \to \mathbb{R}, \quad x \mapsto -\|x\|^2.$$

We define the function $u \in U_{\text{iso},0,0}$ by

$$L(g)u(g) = \begin{pmatrix} 0 & 1 \\ -1 & 0 \end{pmatrix}(g \cdot x_0 - x_0) \qquad \text{for all } g \in \mathcal{G}.$$

We have $y_0 = p \cdot x_0 - x_0 = -2e_1$ and

$$E''(\chi_{\mathcal{G}}x_0)(u,u) = \frac{1}{|\mathcal{G}|} \sum_{g \in \mathcal{G}} V''(y_0)(-u(gp) - u(g), -u(gp) - u(g))$$

$$= V''(y_0)(u(id) + u(p), u(id) + u(p))$$

$$= -2\|u(id) + u(p)\|^2$$

$$= -8.$$

Since $\|u\|_{\mathcal{R}} = \|u\|_{\mathcal{R},0,0} = 0$, we have $\lambda_a = \lambda_{a,0,0} = -\infty$ and $E''(\chi_{\mathcal{G}}x_0)$ is not bounded with respect to $\|\cdot\|_{\mathcal{R},0,0}$.

Example 4.41. We present an example so that $E'(\chi_{\mathcal{G}} x_0) = 0$ and $\lambda_a = \lambda_{a,0,0} = -\infty$. In particular $E''(\chi_{\mathcal{G}} x_0)$ is not bounded with respect to $\|\cdot\|_{\mathcal{R},0,0}$.

Let $d = d_1 = 1$, $d_2 = 0$, $t = (I_1, 1) \in E(1)$, $\mathcal{G} = \{t^n \mid n \in \mathbb{Z}\} < E(1)$ and $x_0 = 0 \in \mathbb{R}$. We have $M_0 = \mathbb{N}$. Let $\alpha > 1$ and $V \colon (\mathbb{R}^d)^{\mathcal{G}\backslash\{id\}} \to \mathbb{R}$ be the interaction potential such that V has the properties (H1), (H2) and (H3) and

$$V''(y_0)(z_1, z_2) = -\sum_{n \in \mathbb{N}} n^{-\alpha} z_1(t^n) z_2(t^n) \text{ for all } z_1, z_2 \in L^\infty(\mathcal{G} \setminus \{id\}, \mathbb{R}^d).$$

We have $E'(\chi_{\mathcal{G}} x_0) = 0$ by Corollary 4.17. Let $N \in \mathbb{N}$ be even. The set $\{t^0, \ldots, t^{N-1}\}$ is a representation set of $\mathcal{G}/T^N$. We define the T^N-periodic function $u \in U_{\text{per}}$ by

$$u(t^n) = \frac{n}{N} \qquad\qquad \text{for all } n \in \{0, \ldots, N/2 - 1\}$$

and

$$u(t^n) = 1 - \frac{n}{N} \qquad\qquad \text{for all } n \in \{N/2, \ldots, N-1\}.$$

Let $\mathcal{R} = \{id, t, t^2\}$ and $\mathcal{R}' = \{t\}$. The set $\mathcal{R}$ has Property 2 and $\mathcal{R}'$ generates $\mathcal{G}$. By Corollary 3.42 and Theorem 3.40, the seminorms $\|\cdot\|_{\mathcal{R}}$ and $\|\nabla_{\mathcal{R}'} \cdot\|$ are equivalent and thus there exists a constant $C > 0$ such that $\|\cdot\|_{\mathcal{R}} \leq C \|\nabla_{\mathcal{R}'} \cdot\|$. We have

$$\|u\|_{\mathcal{R}} \leq C \|\nabla_{\mathcal{R}'} u\|_2 = C \left(\frac{1}{N} \sum_{n=0}^{N-1} \|\nabla_{\mathcal{R}'} u(t^n)\|^2 \right)^{\frac{1}{2}} = \frac{C}{N}. \qquad (4.34)$$

We have

$$E''(\chi_{\mathcal{G}} x_0)(u, u)$$

$$= \frac{1}{N} \sum_{n=0}^{N-1} V''(y_0)\Big(\big(u(t^n s) - u(t^n)\big)_{s \in \mathcal{G}\backslash\{id\}}, \big(u(t^n s) - u(t^n)\big)_{s \in \mathcal{G}\backslash\{id\}} \Big)$$

$$= -\frac{1}{N} \sum_{n=0}^{N-1} \sum_{m \in \mathbb{N}} m^{-\alpha} \big|u(t^{n+m}) - u(t^n)\big|^2$$

$$\leq -\frac{1}{N} \sum_{n=0}^{N-1} (N/2)^{-\alpha} \big|u(t^{n+N/2}) - u(t^n)\big|^2$$

$$\leq -\frac{1}{2}(N/2)^{-\alpha} \frac{1}{4^2}$$

$$= -2^{\alpha-5} N^{-\alpha}. \qquad (4.35)$$

By (4.34) and (4.35) we have

$$\lambda_{\mathrm{a}} \leq \frac{E''(\chi_{\mathcal{G}}x_0)(u,u)}{\|u\|_{\mathcal{R}}^2} \leq -cN^{2-\alpha},$$

where $c = C^{-2}2^{\alpha-5}$. For all $\alpha \in (1,2)$ we have $\lambda_{\mathrm{a}} = -\infty$ since $N \in 2\mathbb{N}$ was arbitrary. Since $\| \cdot \|_{\mathcal{R}} = \| \cdot \|_{\mathcal{R},0,0}$, for all $\alpha \in (1,2)$ we also have $\lambda_{\mathrm{a},0,0} = -\infty$.

4.5. The main theorem

In this section we characterize the stability constants λ_{a} and $\lambda_{\mathrm{a},0,0}$ in the Fourier transform domain, see Theorem 4.51. We also state a similar characterization which enables us to efficiently compute λ_{a} and $\lambda_{\mathrm{a},0,0}$, see Theorem 4.54.

Recall Definition 2.61. Since $E''(\chi_{\mathcal{G}}x_0)$ is left-translation-invariant, see Remark 4.15(i), we can represent $E''(\chi_{\mathcal{G}}x_0)$ as a convolution operator.

Lemma 4.42. *For all $u, v \in U_{\mathrm{per}}$ we have*

$$E''(\chi_{\mathcal{G}}x_0)(u,v) = \langle f_V * v_0, u_0 \rangle,$$

where $u_0 = u(\,\cdot\,^{-1})$ and $v_0 = v(\,\cdot\,^{-1})$.

Proof. Let $u, v \in U_{\mathrm{per}}$. Let $N \in M_0$ such that u and v are $\mathcal{T}^N$-periodic. Let $u_0 = u(\,\cdot\,^{-1})$ and $v_0 = v(\,\cdot\,^{-1})$. By Lemma 4.14 we have

$$\begin{aligned}
E''(\chi_{\mathcal{G}}x_0)(u,v) &= \sum_{g,h \in \mathcal{C}_N} u(g)^{\mathsf{T}} \partial_{g\mathcal{T}^N} \partial_{h\mathcal{T}^N} E''(\chi_{\mathcal{G}}x_0) v(h) \\
&= \frac{1}{|\mathcal{C}_N|} \sum_{g,h \in \mathcal{C}_N} \sum_{t \in \mathcal{T}^N} u_0(g^{-1})^{\mathsf{T}} f_V(g^{-1}ht) v_0(h^{-1}) \\
&= \frac{1}{|\mathcal{C}_N|} \sum_{g \in \mathcal{C}_N} u_0(g^{-1})^{\mathsf{T}} f_V * v_0(g^{-1}) \\
&= \langle f_V * v_0, u_0 \rangle,
\end{aligned}$$

where in the third step we used that $v_0((ht)^{-1}) = v_0(h^{-1})$ for all $h \in \mathcal{C}_N$ and $t \in \mathcal{T}^N$. $\qquad\square$

Let $\varphi \colon \mathcal{R} \to \{0, \ldots, |\mathcal{R}| - 1\}$ be a bijection. We define an isomorphism between $\mathbb{C}^{(m|\mathcal{R}|) \times n}$ and $(\mathbb{C}^{m \times n})^{\mathcal{R}}$ by

$$(a_{i,j})_{i \in \{1,\ldots,m|\mathcal{R}|\}; j \in \{1,\ldots,n\}} \mapsto \left((a_{i+m\varphi(g),j})_{i \in \{1,\ldots,m\}; j \in \{1,\ldots,n\}} \right)_{g \in \mathcal{R}}.$$

Definition 4.43. We define the functions $g_{\mathcal{R}}, g_{\mathcal{R},0,0} \in L^1(\mathcal{G}, \mathbb{R}^{(d|\mathcal{R}|)\times d})$ by

$$g_{\mathcal{R}}(g) = P(\delta_{g,h}I_d)_{h\in\mathcal{R}} \qquad\qquad \text{for all } g \in \mathcal{G}$$

and

$$g_{\mathcal{R},0,0}(g) = P_0(\delta_{g,h}I_d)_{h\in\mathcal{R}} \qquad\qquad \text{for all } g \in \mathcal{G},$$

where P (resp. P_0) is the square matrix of order $d|\mathcal{R}|$ such that the map

$$\mathbb{R}^{d|\mathcal{R}|} \to \mathbb{R}^{d|\mathcal{R}|}, \quad x \mapsto Px$$

is the orthogonal projection with respect to the norm $\|\cdot\|$ with kernel $U_{\text{iso}}(\mathcal{R})$ (resp. $U_{\text{iso},0,0}(\mathcal{R})$).

Remark 4.44. The support of both $g_{\mathcal{R}}$ and $g_{\mathcal{R},0,0}$ is equal to $\mathcal{R}$. We have

$$g_{\mathcal{R}}(g) = p_{\varphi(g)} \qquad\qquad \text{for all } g \in \mathcal{R}$$

and

$$g_{\mathcal{R},0,0}(g) = p_{0,\varphi(g)} \qquad\qquad \text{for all } g \in \mathcal{R},$$

where $p_0, \ldots, p_{|\mathcal{R}|-1}, p_{0,0}, \ldots, p_{0,|\mathcal{R}|-1} \in \mathbb{R}^{(d|\mathcal{R}|)\times d}$ such that $P = (p_0, \ldots, p_{|\mathcal{R}|-1})$ and $P_0 = (p_{0,0}, \ldots, p_{0,|\mathcal{R}|-1})$ and both P and P_0 are as above.

Due to the left-translation-invariance, $\|\cdot\|_{\mathcal{R}}$ and $\|\cdot\|_{\mathcal{R},0,0}$ can be represented by means of convolution operators.

Lemma 4.45. *For all $u \in U_{\text{per}}$ we have that $\|u\|_{\mathcal{R}} = \|g_{\mathcal{R}} * u_0\|_2$ and $\|u\|_{\mathcal{R},0,0} = \|g_{\mathcal{R},0,0} * u_0\|_2$, where $u_0 = u(\,\cdot\,^{-1})$.*

Proof. Let $u \in U_{\text{per}}$ and $N \in M_0$ such that u is $\mathcal{T}^N$-periodic. Let $u_0 = u(\,\cdot\,^{-1})$. Let $P \in \mathbb{R}^{(d|\mathcal{R}|)\times(d|\mathcal{R}|)}$ such that the map

$$\mathbb{R}^{d|\mathcal{R}|} \to \mathbb{R}^{d|\mathcal{R}|}, x \mapsto Px$$

is the orthogonal projection with kernel $U_{\text{iso}}(\mathcal{R})$. We have

$$\|u\|_{\mathcal{R}}^2 = \frac{1}{|\mathcal{C}_N|} \sum_{g\in\mathcal{C}_N} \|P(u(gh))_{h\in\mathcal{R}}\|^2. \tag{4.36}$$

For all $g \in \mathcal{G}$ we define the function

$$\delta_g \colon \mathcal{G} \to \{0,1\}, h \mapsto \delta_{h,g}.$$

For all $g \in \mathcal{G}$ we have

$$
\begin{aligned}
P(u(gh))_{h \in \mathcal{R}} &= P(u_0(h^{-1}g^{-1}))_{h \in \mathcal{R}} \\
&= P((\delta_h I_d) * u_0(g^{-1}))_{h \in \mathcal{R}} \\
&= (P(\delta_h I_d)_{h \in \mathcal{R}}) * u_0(g^{-1}) \\
&= g_{\mathcal{R}} * u_0(g^{-1}).
\end{aligned} \tag{4.37}
$$

By (4.36) and (4.37) we have

$$
\|u\|_{\mathcal{R}}^2 = \frac{1}{|\mathcal{C}_N|} \sum_{g \in \mathcal{C}_N} \|g_{\mathcal{R}} * u_0(g^{-1})\|^2 = \|g_{\mathcal{R}} * u_0\|_2^2.
$$

Analogously we have $\|u\|_{\mathcal{R},0,0} = \|g_{\mathcal{R},0,0} * u_0\|_2$. $\qquad \square$

Proposition 4.29 implies the following corollary.

Corollary 4.46. *Suppose that $E'(\chi_{\mathcal{G}} x_0) = 0$. Then for all periodic representations ρ of $\mathcal{G}$ and $a \in \mathbb{C}^{dd_\rho}$ such that $\|\widehat{g_{\mathcal{R}}}(\rho)a\| = 0$ we have $\langle \widehat{f_V}(\rho)a, a \rangle = 0$.*

Proof. Suppose that $E'(\chi_{\mathcal{G}} x_0) = 0$. Let ρ be a periodic representation of $\mathcal{G}$ and $a \in \mathbb{C}^{dd_\rho}$ such that $\|\widehat{g_{\mathcal{R}}}(\rho)a\| = 0$. Without loss of generality we assume that $\rho \in \mathcal{E}$. We define $u \in U_{\mathrm{per},\mathbb{C}}$ by

$$
\widehat{u}(\rho') = \begin{cases} (a \quad 0_{dd_\rho, d_\rho - 1}) & \text{if } \rho' = \rho \\ 0_{dd_{\rho'}, d_{\bar\rho}} & \text{else} \end{cases}
$$

for all $\rho' \in \mathcal{E}$. We denote $u_0 = u(\cdot^{-1})$. We have

$$
\begin{aligned}
0 = d_\rho \|\widehat{g_{\mathcal{R}}}(\rho)a\|^2 &= d_\rho \|\widehat{g_{\mathcal{R}}}(\rho)\widehat{u}(\rho)\|^2 = d_\rho \|\widehat{g_{\mathcal{R}} * u}(\rho)\|^2 = \|g_{\mathcal{R}} * u\|_2^2 \\
&= \|g_{\mathcal{R}} * \mathrm{Re}(u)\|_2^2 + \|g_{\mathcal{R}} * \mathrm{Im}(u)\|_2^2 = \|\mathrm{Re}(u_0)\|_{\mathcal{R}}^2 + \|\mathrm{Im}(u_0)\|_{\mathcal{R}}^2, \tag{4.38}
\end{aligned}
$$

where we used Proposition 2.56 in the third step and Lemma 4.45 in the last step. Thus we have $\|\mathrm{Re}(u_0)\|_{\mathcal{R}} = 0$ and $\|\mathrm{Im}(u_0)\|_{\mathcal{R}} = 0$ which is equivalent to $\mathrm{Re}(u_0), \mathrm{Im}(u_0) \in U_{\mathrm{iso},0,0}$ by Theorem 3.34. We have $E''(\chi_{\mathcal{G}} x_0)(\mathrm{Re}(u_0), \mathrm{Re}(u_0)) = 0$ and $E''(\chi_{\mathcal{G}} x_0)(\mathrm{Im}(u_0), \mathrm{Im}(u_0)) = 0$ by Proposition 4.29 and Remark 4.30(ii). Thus we have

$$
\begin{aligned}
d_\rho \langle \widehat{f_V}(\rho)a, a \rangle &= E''(\chi_{\mathcal{G}} x_0)(\mathrm{Re}(u_0), \mathrm{Re}(u_0)) + E''(\chi_{\mathcal{G}} x_0)(\mathrm{Im}(u_0), \mathrm{Im}(u_0)) \\
&= 0,
\end{aligned}
$$

where the first step follows analogously to (4.38) with Lemma 4.42 instead of Lemma 4.45. $\qquad \square$

The following lemma shows that we can consider complex-valued instead of real-valued functions.

Lemma 4.47. *We have*

$$\lambda_a = \sup\big\{c \in \mathbb{R} \,\big|\, \forall u \in U_{\mathrm{per},\mathbb{C}} : c\|g_{\mathcal{R}} * u\|_2^2 \le \langle f_V * u, u\rangle\big\}$$

and

$$\lambda_{a,0,0} = \sup\big\{c \in \mathbb{R} \,\big|\, \forall u \in U_{\mathrm{per},\mathbb{C}} : c\|g_{\mathcal{R},0,0} * u\|_2^2 \le \langle f_V * u, u\rangle\big\}.$$

Proof. By Lemma 4.42, Lemma 4.45 and since $U_{\mathrm{per}} = \{u(\,\cdot^{-1})\,|\,u \in U_{\mathrm{per}}\}$, we have

$$\lambda_{\mathrm{a}} = \sup\big\{c \in \mathbb{R} \,\big|\, \forall u \in U_{\mathrm{per}} : c\|g_{\mathcal{R}} * u\|_2^2 \le \langle f_V * u, u\rangle\big\}$$

and hence,

$$\lambda_{\mathrm{a}} \ge \sup\big\{c \in \mathbb{R} \,\big|\, \forall u \in U_{\mathrm{per},\mathbb{C}} : c\|g_{\mathcal{R}} * u\|_2^2 \le \langle f_V * u, u\rangle\big\} =: \mathrm{RHS}.$$

Now we show that $\lambda_{\mathrm{a}} \le \mathrm{RHS}$. For all $u \in U_{\mathrm{per},\mathbb{C}}$ we have

$$
\begin{aligned}
\langle f_V * u, u\rangle &= \langle f_V * \mathrm{Re}(u), \mathrm{Re}(u)\rangle - \mathrm{i}\langle f_V * \mathrm{Re}(u), \mathrm{Im}(u)\rangle \\
&\quad + \mathrm{i}\langle f_V * \mathrm{Im}(u), \mathrm{Re}(u)\rangle + \langle f_V * \mathrm{Im}(u), \mathrm{Im}(u)\rangle \\
&= \langle f_V * \mathrm{Re}(u), \mathrm{Re}(u)\rangle - \mathrm{i}E''(\chi_{\mathcal{G}}x_0)(\mathrm{Im}(u), \mathrm{Re}(u)) \\
&\quad + \mathrm{i}E''(\chi_{\mathcal{G}}x_0)(\mathrm{Re}(u), \mathrm{Im}(u)) + \langle f_V * \mathrm{Im}(u), \mathrm{Im}(u)\rangle \\
&= \langle f_V * \mathrm{Re}(u), \mathrm{Re}(u)\rangle + \langle f_V * \mathrm{Im}(u), \mathrm{Im}(u)\rangle \\
&\ge \lambda_{\mathrm{a}}\|g_{\mathcal{R}} * \mathrm{Re}(u)\|_2^2 + \lambda_{\mathrm{a}}\|g_{\mathcal{R}} * \mathrm{Im}(u)\|_2^2 \\
&= \lambda_{\mathrm{a}}\|g_{\mathcal{R}} * u\|_2^2,
\end{aligned}
$$

where in the second step we used Lemma 4.42.
The proof of the characterization of $\lambda_{\mathrm{a},0,0}$ is analogous. $\square$

Recall that by Definition 2.19 all representations are unitary. Schwarz's theorem implies the following lemma.

Lemma 4.48. *For all $g \in \mathcal{G}$ we have $f_V(g^{-1}) = f_V(g)^{\mathsf{T}}$ and for all representations ρ of $\mathcal{G}$ the matrix $\widehat{f_V}(\rho)$ is Hermitian.*

Proof. For all $g \in \mathcal{G}$ we have

$$
\begin{aligned}
f_V(g^{-1}) = \sum_{h_1,h_2 \in \mathcal{G}\backslash\{id\}} \Big(&\delta_{g^{-1},h_2^{-1}h_1} L(h_2)^{\mathsf{T}}\partial_{h_2}\partial_{h_1}V(y_0)L(h_1) \\
&- \delta_{g^{-1},h_2^{-1}} L(h_2)^{\mathsf{T}}\partial_{h_2}\partial_{h_1}V(y_0) - \delta_{g^{-1},h_1}\partial_{h_2}\partial_{h_1}V(y_0)L(h_1) \\
&+ \delta_{g^{-1},id}\partial_{h_2,h_1}V(y_0)\Big)
\end{aligned}
$$

$$= \sum_{h_1,h_2\in\mathcal{G}\setminus\{id\}} \Big(\delta_{g,h_1^{-1}h_2} L(h_2)^{\mathsf{T}}(\partial_{h_1}\partial_{h_2}V(y_0))^{\mathsf{T}} L(h_1)$$
$$- \delta_{g,h_2} L(h_2)^{\mathsf{T}}(\partial_{h_1}\partial_{h_2}V(y_0))^{\mathsf{T}}$$
$$- \delta_{g,h_1^{-1}}(\partial_{h_1}\partial_{h_2}V(y_0))^{\mathsf{T}}L(h_1) + \delta_{g,id}(\partial_{h_1}\partial_{h_2}V(y_0))^{\mathsf{T}} \Big)$$
$$= f_V(g)^{\mathsf{T}}. \tag{4.39}$$

For all representations ρ of $\mathcal{G}$ we have

$$\widehat{f_V}(\rho) = \sum_{g\in\mathcal{G}} f_V(g)\otimes\rho(g)$$
$$= \sum_{g\in\mathcal{G}} f_V(g^{-1})\otimes\rho(g^{-1})$$
$$= \sum_{g\in\mathcal{G}} f_V(g)^{\mathsf{H}}\otimes\rho(g)^{\mathsf{H}}$$
$$= \Big(\sum_{g\in\mathcal{G}} f_V(g)\otimes\rho(g)\Big)^{\mathsf{H}}$$
$$= \widehat{f_V}(\rho)^{\mathsf{H}},$$

where in the third step we used (4.39) and that ρ is unitary. $\qquad\square$

Definition 4.49. The *Loewner order* is the partial order on the set of all Hermitian matrices of $\mathbb{C}^{n\times n}$ defined by $A \geq B$ if $A - B$ is positive semidefinite. We define

$$\lambda_{\min}(A,B) := \sup\{c\in\mathbb{R} \mid cB^{\mathsf{H}}B \leq A\} \in \mathbb{R}\cup\{\pm\infty\}$$

for all Hermitian matrices $A\in\mathbb{C}^{n\times n}$ and matrices $B\in\mathbb{C}^{m\times n}$.

Remark 4.50.   (i) By means of the dual problem we have

$$\lambda_{\min}(A,B) = \inf\{x^{\mathsf{H}}Ax \mid x\in\mathbb{C}^n, \|Bx\| = 1\}$$

and

$$\lambda_{\min}(A,0_{m,n}) = \begin{cases} \infty & \text{if } A \text{ is positive semidefinite} \\ -\infty & \text{else} \end{cases}$$

for all Hermitian matrices $A\in\mathbb{C}^{n\times n}$ and matrices $B\in\mathbb{C}^{m\times n}\setminus\{0\}$. The proof is analogous to the proof of Proposition 4.10.

(ii) Suppose that B has in addition rank n and consider the *generalized* eigenvalue problem $Av = \lambda B^H Bv$, i.e. the problem of finding the *eigenvalues* of the *matrix pencil* $A - \lambda B^H B$. Then the eigenvalues of the generalized eigenvalue problem are real and $\lambda_{\min}(A, B)$ is equal to the smallest eigenvalue of the generalized eigenvalue problem, see [34, Chapter X, Theorem 11]. The eigenvalues of the generalized eigenvalue problem are equal to the eigenvalues of the matrix $A(B^H B)^{-1}$, see [58, Proposition 6.1.1], but the eigenvalues of $A(B^H B)^{-1}$ are ill-conditioned. There exist many numerically stable algorithms, see, e.g., [8, Chapter 5], and thus many programming languages have a function for this problem; e.g. for Python the subpackage linalg of the package SciPy has the function eigvalsh.

Due to the left-translation-invariance, $E''(\chi_\mathcal{G} x_0)$, $\| \cdot \|_\mathcal{R}$ and $\| \cdot \|_{\mathcal{R},0,0}$ can be represented by means of multiplier operators. Thus we have the following representation of λ_a and $\lambda_{a,0,0}$. Recall that $\mathcal{E}$ is a representation set of $\{\rho \in \widehat{\mathcal{G}} \,|\, \rho \text{ is periodic}\}$.

Theorem 4.51. *We have*

$$\lambda_a = \inf\left\{\lambda_{\min}\left(\widehat{f_V}(\rho), \widehat{g_\mathcal{R}}(\rho)\right) \,\Big|\, \rho \in \mathcal{E}\right\}$$

and

$$\lambda_{a,0,0} = \inf\left\{\lambda_{\min}\left(\widehat{f_V}(\rho), \widehat{g_{\mathcal{R},0,0}}(\rho)\right) \,\Big|\, \rho \in \mathcal{E}\right\}.$$

Proof. By Lemma 4.48 for all $\rho \in \mathcal{E}$ the matrix $\widehat{f_V}(\rho)$ is Hermitian and thus the term $\lambda_{\min}(\widehat{f_V}(\rho), \widehat{g_\mathcal{R}}(\rho))$ is well-defined. We have to show that

$$\lambda_a = \inf\left\{\lambda_{\min}\left(\widehat{f_V}(\rho), \widehat{g_\mathcal{R}}(\rho)\right) \,\Big|\, \rho \in \mathcal{E}\right\} =: \text{RHS}.$$

By Lemma 4.47 we have

$$\lambda_a = \sup\left\{c \in \mathbb{R} \,\Big|\, \forall u \in U_{\text{per},\mathbb{C}} : c\|g_\mathcal{R} * u\|_2^2 \le \langle f_V * u, u\rangle\right\}.$$

First we show that $\lambda_a \le \text{RHS}$. Let $\rho \in \mathcal{E}$ and $a \in \mathbb{C}^{dd_\rho}$. We define $u \in U_{\text{per},\mathbb{C}}$ by

$$\widehat{u}(\rho') = \begin{cases} (a \quad 0_{dd_\rho, d_\rho - 1}) & \text{if } \rho' = \rho \\ 0_{dd_{\rho'}, d_{\rho'}} & \text{else} \end{cases}$$

for all $\rho' \in \mathcal{E}$. By Lemma 2.62 and Proposition 2.56 we have

$$\left\langle \widehat{f_V}(\rho)a, a\right\rangle = \left\langle \widehat{f_V}(\rho)\widehat{u}(\rho), \widehat{u}(\rho)\right\rangle = \left\langle \widehat{f_V * u}(\rho), \widehat{u}(\rho)\right\rangle = \frac{1}{d_\rho}\langle f_V * u, u\rangle$$

$$\ge \frac{\lambda_a}{d_\rho}\|g_\mathcal{R} * u\|_2^2 = \lambda_a \|\widehat{g_\mathcal{R} * u}(\rho)\|^2 = \lambda_a\|\widehat{g_\mathcal{R}}(\rho)\widehat{u}(\rho)\|^2 = \lambda_a\|\widehat{g_\mathcal{R}}(\rho)a\|^2.$$

Since $a \in \mathbb{C}^{dd_\rho}$ was arbitrary, we have $\lambda_{\min}(\widehat{f_V}(\rho), \widehat{g_\mathcal{R}}(\rho)) \geq \lambda_a$.

Now we prove that $\lambda_a \geq \text{RHS}$. Let $u \in U_{\text{per},\mathbb{C}}$. For a matrix A we denote its ith column by A_i. We have

$$
\begin{aligned}
\langle f_V * u, u \rangle &= \sum_{\rho \in \mathcal{E}} d_\rho \Big\langle \widehat{f_V * u}(\rho), \widehat{u}(\rho) \Big\rangle \\
&= \sum_{\rho \in \mathcal{E}} d_\rho \Big\langle \widehat{f_V}(\rho)\widehat{u}(\rho), \widehat{u}(\rho) \Big\rangle \\
&= \sum_{\rho \in \mathcal{E}} d_\rho \sum_{i=1}^{d_\rho} \Big\langle \widehat{f_V}(\rho)\widehat{u}(\rho)_i, \widehat{u}(\rho)_i \Big\rangle \\
&\geq \text{RHS} \sum_{\rho \in \mathcal{E}} d_\rho \sum_{i=1}^{d_\rho} \|\widehat{g_\mathcal{R}}(\rho)\widehat{u}(\rho)_i\|^2 \\
&= \text{RHS} \sum_{\rho \in \mathcal{E}} d_\rho \|\widehat{g_\mathcal{R}}(\rho)\widehat{u}(\rho)\|^2 \\
&= \text{RHS} \|g_\mathcal{R} * u\|_2^2.
\end{aligned}
$$

The proof of the characterization of $\lambda_{a,0,0}$ is analogous. $\qquad\square$

For the remainder of this section, we fix a complete set of representatives of the cosets of $\mathcal{TF}$ in $\mathcal{G}$ such that $\text{Ind}\,\rho$ is well-defined for all ρ by Definition 2.23. In the following we write $\text{Ind}\,\rho$ for $\text{Ind}_{\mathcal{TF}}^{\mathcal{G}}\,\rho$ for all representations ρ of $\mathcal{TF}$.

Lemma 4.52. *For all representations ρ of $\mathcal{TF}$ the functions*

$$
\mathbb{R}^{d_2} \to \mathbb{C}^{(dd_\rho) \times (dd_\rho)}, \qquad k \mapsto \widehat{f_V}(\text{Ind}(\chi_k \rho)),
$$
$$
\mathbb{R}^{d_2} \to \mathbb{C}^{(|\mathcal{R}|d_\rho) \times (dd_\rho)}, \qquad k \mapsto \widehat{g_\mathcal{R}}(\text{Ind}(\chi_k \rho))
$$

and

$$
\mathbb{R}^{d_2} \to \mathbb{C}^{(|\mathcal{R}|d_\rho) \times (dd_\rho)}, \qquad k \mapsto \widehat{g_{\mathcal{R},0,0}}(\text{Ind}(\chi_k \rho))
$$

are continuous and the functions

$$
\mathbb{R}^{d_2} \to \mathbb{R} \cup \{\pm\infty\}, \qquad k \mapsto \lambda_{\min}\Big(\widehat{f_V}(\text{Ind}(\chi_k \rho)), \widehat{g_\mathcal{R}}(\text{Ind}(\chi_k \rho))\Big)
$$

and

$$
\mathbb{R}^{d_2} \to \mathbb{R} \cup \{\pm\infty\}, \qquad k \mapsto \lambda_{\min}\Big(\widehat{f_V}(\text{Ind}(\chi_k \rho)), \widehat{g_{\mathcal{R},0,0}}(\text{Ind}(\chi_k \rho))\Big)
$$

are upper semicontinuous.

Proof. Let ρ be a representation of $\mathcal{TF}$ and f_i denote the ith function of the lemma for every $i \in \{1, \ldots, 5\}$. By Theorem D.7 the functions f_1, f_2 and f_3 are continuous.

Let $(k_n)_{n \in \mathbb{N}}$ be a sequence in $\mathbb{R}^{d_2}$ and $k \in \mathbb{R}^{d_2}$ be such that $\lim_{n \to \infty} k_n = k$. Without loss of generality we assume that $\limsup_{n \to \infty} f_4(k_n) > -\infty$ and $\limsup_{n \to \infty} f_4(k_n) = \lim_{n \to \infty} f_4(k_n)$. Let $\lambda \in \mathbb{R}$ be such that $\lambda < \limsup_{n \to \infty} f_4(k_n)$. We have $\lambda f_2(k_n)^{\mathsf{H}} f_2(k_n) \leq f_1(k_n)$ for all $n \in \mathbb{N}$ large enough. Since the Loewner order is closed, i.e. the set $\{(A, B) \in X^2 \,|\, A \leq B\}$ is closed, where $X = \{A \in \mathbb{C}^{(dd_\rho) \times (dd_\rho)} \,|\, A \text{ is Hermitian}\}$, we have $\lambda f_2(k)^{\mathsf{H}} f_2(k) \leq f_1(k)$. Thus we have $\lambda \leq f_4(k)$.

Analogously the function f_5 is upper semicontinuous. $\qquad\square$

Recall Definition 2.38, Proposition 2.39 and Definition 2.32.

Definition 4.53. For all $\rho \in \widehat{\mathcal{TF}}$ and representations $\rho' \in \rho$ we define the space group

$$\mathcal{G}_{\rho'} := \mathcal{G}_\rho.$$

The following theorem generalizes Theorem 3.6(b) of [40] from lattices to general configurations.

Theorem 4.54. *Let R be a representation set of a representation set of $\widehat{\mathcal{TF}}/\sim$. For all $\rho \in R$ let K_ρ be a representation set of $\mathbb{R}^{d_2}/\mathcal{G}_\rho$. Then we have*

$$\lambda_a = \inf\left\{ \lambda_{\min}\left(\widehat{f_V}(\mathrm{Ind}(\chi_k \rho)), \widehat{g_{\mathcal{R}}}(\mathrm{Ind}(\chi_k \rho)) \right) \,\middle|\, \rho \in R, k \in K_\rho \right\}$$

and

$$\lambda_{a,0,0} = \inf\left\{ \lambda_{\min}\left(\widehat{f_V}(\mathrm{Ind}(\chi_k \rho)), \widehat{g_{\mathcal{R},0,0}}(\mathrm{Ind}(\chi_k \rho)) \right) \,\middle|\, \rho \in R, k \in K_\rho \right\}.$$

Proof. Let R be a representation set of a representation set of $\widehat{\mathcal{TF}}/\sim$. For all $\rho \in R$ let K_ρ be a representation set of $\mathbb{R}^{d_2}/\mathcal{G}_\rho$. Let $m \in \mathbb{N}$ such that $M_0 = m\mathbb{N}$. By Lemma 2.36(i) there exists a representation set R' of a representation set of $\widehat{\mathcal{TF}}/\sim$ such that ρ is $\mathcal{T}^m$-periodic for all $\rho \in R'$. Due to the existence of fundamental domains, see, e. g., [49, Theorem 6.6.13], for all $\rho \in R'$ there exists a representation set K'_ρ of $\mathbb{R}^{d_2}/\mathcal{G}_\rho$ such that L'_ρ is a dense subset of K'_ρ, where $L'_\rho = \{k \in K'_\rho \,|\, \exists N \in M_0 : k \in L^*_{\mathcal{S}}/N\}$. By Theorem 2.43 applied to R and R', there exist a bijection

$$\varphi \colon \bigsqcup_{\rho \in R'} K'_\rho \to \bigsqcup_{\rho \in R} K_\rho, \quad (k, \rho) \mapsto (\varphi_1(k, \rho), \varphi_2(k, \rho)) \tag{4.40}$$

and for all $\rho \in R'$ and $k \in K'_\rho$ some $T_{k,\rho} \in \mathrm{U}(d_{\mathrm{Ind}(\chi_k\rho)})$ such that

$$\mathrm{Ind}(\chi_{\varphi_1(k,\rho)}\varphi_2(k,\rho)) = T^{\mathsf{H}}_{k,\rho}\,\mathrm{Ind}(\chi_k\rho)T_{k,\rho}. \tag{4.41}$$

By (4.40) and (4.41) we have

$$
\begin{aligned}
\mathrm{RHS} &:= \inf\Big\{\lambda_{\min}\Big(\widehat{f_V}(\mathrm{Ind}(\chi_k\rho)),\widehat{g_{\mathcal{R}}}(\mathrm{Ind}(\chi_k\rho))\Big)\,\Big|\,\rho \in R, k \in K_\rho\Big\}\\
&= \inf\Big\{\lambda_{\min}\Big(\widehat{f_V}\big(\mathrm{Ind}\big(\chi_{\varphi_1(k,\rho)}\varphi_2(k,\rho)\big)\big),\\
&\qquad\qquad \widehat{g_{\mathcal{R}}}\big(\mathrm{Ind}\big(\chi_{\varphi_1(k,\rho)}\varphi_2(k,\rho)\big)\big)\Big)\,\Big|\,\rho \in R', k \in K'_\rho\Big\}\\
&= \inf\Big\{\lambda_{\min}\Big(\big(I_d \otimes T^{\mathsf{H}}_{k,\rho}\big)\widehat{f_V}(\mathrm{Ind}(\chi_k\rho))\big(I_d \otimes T_{k,\rho}\big),\\
&\qquad\qquad \big(I_d \otimes T^{\mathsf{H}}_{k,\rho}\big)\widehat{g_{\mathcal{R}}}(\mathrm{Ind}(\chi_k\rho))\big(I_d \otimes T_{k,\rho}\big)\Big)\,\Big|\,\rho \in R', k \in K'_\rho\Big\}\\
&= \inf\Big\{\lambda_{\min}\Big(\widehat{f_V}(\mathrm{Ind}(\chi_k\rho)),\widehat{g_{\mathcal{R}}}(\mathrm{Ind}(\chi_k\rho))\Big)\,\Big|\,\rho \in R', k \in K'_\rho\Big\}.
\end{aligned}
\tag{4.42}
$$

For all $\rho \in R'$ we define the function

$$
\begin{aligned}
f_\rho\colon K'_\rho &\to \mathbb{R} \cup \{\pm\infty\}\\
k &\mapsto \lambda_{\min}\Big(\widehat{f_V}(\mathrm{Ind}(\chi_k\rho)),\widehat{g_{\mathcal{R}}}(\mathrm{Ind}(\chi_k\rho))\Big).
\end{aligned}
$$

By Lemma 4.52 for all $\rho \in R$ the function f_ρ is upper semicontinuous and thus we have

$$\inf f_\rho = \inf f_\rho|_{L'_\rho}. \tag{4.43}$$

By (4.42) and (4.43) we have

$$\mathrm{RHS} = \inf\big\{f_\rho(k)\,\big|\,\rho \in R', k \in L'_\rho\big\}.$$

By Theorem 4.51 we have

$$\lambda_{\mathrm{a}} = \inf\Big\{\lambda_{\min}\Big(\widehat{f_V}(\rho),\widehat{g_{\mathcal{R}}}(\rho)\Big)\,\Big|\,\rho \in \mathcal{E}\Big\}. \tag{4.44}$$

By Lemma D.3(ii) there exists a permutation matrix $P_{n,p_1,\ldots,p_k} \in \mathrm{O}(n(p_1 + \cdots + p_k))$ for all $n, p_1, \ldots, p_k \in \mathbb{N}$ such that

$$A \otimes (B_1 \oplus \cdots \oplus B_k) = P^{\mathsf{T}}_{m,p_1,\ldots,p_k}((A \otimes B_1) \oplus \cdots \oplus (A \otimes B_k))P_{n,p_1,\ldots,p_k}$$

for all $A \in \mathbb{C}^{m \times n}$ and $B_i \in \mathbb{C}^{p_i \times p_i}$, $i \in \{1, \ldots, k\}$.

Now we show that $\lambda_{\mathrm{a}} \leq$ RHS. Let $\rho \in R'$, $k \in L'_\rho$ and $\rho' = \mathrm{Ind}(\chi_k\rho)$. Let $N \in M_0$ such that by Lemma 2.31 and the construction of L'_ρ we have $\chi_k|_{\mathcal{T}^N} = 1$. The map ρ' is $\mathcal{T}^N$-periodic. There exist some $\rho_1, \ldots, \rho_n \in \mathcal{E}$ and $T \in \mathrm{U}(d_{\rho'})$ such that

$$\rho'(g) = T^{\mathsf{H}}(\rho_1(g) \oplus \cdots \oplus \rho_n(g))T \qquad \text{for all } g \in \mathcal{G}.$$

We have

$$
\begin{aligned}
\widehat{f_V}(\rho') &= \sum_{g \in \mathcal{G}} f_V(g) \otimes \rho'(g) \\
&= \sum_{g \in \mathcal{G}} f_V(g) \otimes \big(T^{\mathsf{H}}(\rho_1(g) \oplus \cdots \oplus \rho_n(g))T\big) \\
&= (I_d \otimes T)^{\mathsf{H}}\bigg(\sum_{g \in \mathcal{G}} f_V(g) \otimes \big(\rho_1(g) \oplus \cdots \oplus \rho_n(g)\big)\bigg)(I_d \otimes T) \\
&= P^{\mathsf{H}}\bigg(\bigg(\sum_{g \in \mathcal{G}} f_V(g) \otimes \rho_1(g)\bigg) \oplus \cdots \oplus \bigg(\sum_{g \in \mathcal{G}} f_V(g) \otimes \rho_1(g)\bigg)\bigg)P \\
&= P^{\mathsf{H}}\big(\widehat{f_V}(\rho_1) \oplus \cdots \oplus \widehat{f_V}(\rho_n)\big)P, \qquad\qquad (4.45)
\end{aligned}
$$

where P is the unitary matrix $P_{d,d_{\rho_1},\ldots,d_{\rho_n}}(I_d \otimes T)$. Analogously to (4.45) we have

$$\widehat{g_{\mathcal{R}}}(\rho') = Q^{\mathsf{H}}\big(\widehat{g_{\mathcal{R}}}(\rho_1) \oplus \cdots \oplus \widehat{g_{\mathcal{R}}}(\rho_n)\big)P, \qquad\qquad (4.46)$$

where Q is the unitary matrix $P_{d|\mathcal{R}|,d_{\rho_1},\ldots,d_{\rho_n}}(I_{d|\mathcal{R}|} \otimes T)$. By (4.45), (4.46) and (4.44) we have

$$
\begin{aligned}
f_\rho(k) &= \lambda_{\min}\big(\widehat{f_V}(\rho_1) \oplus \cdots \oplus \widehat{f_V}(\rho_n), \widehat{g_{\mathcal{R}}}(\rho_1) \oplus \cdots \oplus \widehat{g_{\mathcal{R}}}(\rho_n)\big) \\
&= \min\Big\{\lambda_{\min}\big(\widehat{f_V}(\rho_i), \widehat{g_{\mathcal{R}}}(\rho_i)\big) \,\Big|\, i \in \{1, \ldots, n\}\Big\} \\
&\geq \lambda_{\mathrm{a}}.
\end{aligned}
$$

Now we show that $\lambda_{\mathrm{a}} \geq$ RHS. Let $\rho_1 \in \mathcal{E}$. By Corollary 2.44(i) the set $\{\mathrm{Ind}(\chi_k\rho) \,|\, \rho \in R', k \in L'_\rho\}$ is a representation set of $\mathrm{Ind}(\{\rho \in \widehat{\mathcal{TF}} \,|\, \rho$ is periodic$\})$. By [28, p. 1248] there exist some $\rho \in R'$ and $k \in L'_\rho$ such that ρ_1 is isomorphic to a subrepresentation of $\mathrm{Ind}(\chi_k\rho)$. Let $\rho' = \mathrm{Ind}(\chi_k\rho)$. There exist some $\rho_2, \ldots, \rho_n \in \mathcal{E}$ and $T \in \mathrm{U}(d_{\rho'})$ such that

$$\rho'(g) = T^{\mathsf{H}}(\rho_1(g) \oplus \cdots \oplus \rho_n(g))T \qquad \text{for all } g \in \mathcal{G}.$$

Analogously to (4.45) and (4.46) we have

$$\widehat{f_V}(\rho') = P^{\mathsf{H}}\Big(\widehat{f_V}(\rho_1) \oplus \cdots \oplus \widehat{f_V}(\rho_n)\Big)P$$

and

$$\widehat{g_{\mathcal{R}}}(\rho') = Q^{\mathsf{H}}\Big(\widehat{g_{\mathcal{R}}}(\rho_1) \oplus \cdots \oplus \widehat{g_{\mathcal{R}}}(\rho_n)\Big)P,$$

where P is the unitary matrix $P_{d,d_{\rho_1},\dots,d_{\rho_n}}(I_d \otimes T)$ and Q is the unitary matrix $P_{d|\mathcal{R}|,d_{\rho_1},\dots,d_{\rho_n}}(I_{d|\mathcal{R}|} \otimes T)$. We have

$$\lambda_{\min}\Big(\widehat{f_V}(\rho_1),\widehat{g_{\mathcal{R}}}(\rho_1)\Big) \geq \min\Big\{\lambda_{\min}\Big(\widehat{f_V}(\rho_i),\widehat{g_{\mathcal{R}}}(\rho_i)\Big)\,\Big|\, i \in \{1,\dots,n\}\Big\}$$
$$= f_\rho(k) \geq \text{RHS}.$$

The proof of the characterization of $\lambda_{\mathrm{a},0,0}$ is analogous. $\qquad\square$

Remark 4.55. (i) By Lemma 4.52 the above theorem is also true if for all $\rho \in R$ we weaken the assumption on K_ρ and only assume that the closure of K_ρ contains a representation set of $\mathbb{R}^{d_2}/\mathcal{G}_\rho$. In particular the theorem is also true if for all $\rho \in R$ the set K_ρ is a fundamental domain of $\mathbb{R}^{d_2}/K_\rho$.

(ii) An algorithm for the determination of a representation set of $\widehat{\mathcal{TF}}/\sim$ with the aid of the finite group $(\mathcal{TF})_m$ is given by Lemma 2.36, where $m \in \mathbb{N}$ such that $M_0 = m\mathbb{N}$.

4.6. An algorithm to check stability

Due to the main results of the thesis, we can now give an algorithm which checks if $(\mathcal{G}, x_0, V)$ is stable with respect to $\|\cdot\|_{\mathcal{R}}$, see Definition 4.8. The algorithm for the stability with respect to $\|\cdot\|_{\mathcal{R},0,0}$ is analogous.

Algorithm 4.56. Given is a discrete group $\mathcal{G} < \mathrm{E}(d)$ and its associated groups $\mathcal{F}$, $\mathcal{S}$ and set $\mathcal{T}$, see Definition 2.6, some point $x_0 \in \mathbb{R}^d$ such that the map $\mathcal{G} \to \mathbb{R}^d$, $g \mapsto g \cdot x_0$ is injective, and an interaction potential V, see Definition 4.1. Since the algorithm is numeric and by (H3), we may assume that V has finite support.

(i) Check if $\chi_{\mathcal{G}x_0}$ is a critical point of the configurational energy E, e. g. by computing the derivative $\partial_g V(y_0)$ for all $g \in \mathrm{supp}\, V$, see Definition 4.1, the vector e_V, see Definition 4.11 and checking if $e_V = 0$, see Corollary 4.16.

(ii) Determine the derivative $\partial_g \partial_h V(y_0)$ for all $g, h \in \operatorname{supp} V$, see Definition 4.1. Then compute the function f_V by computing $f_V(g)$ for all $g \in (\{id\} \cup \operatorname{supp} V)^{-1}(\{id\} \cup \operatorname{supp} V)$, see Definition 4.11 and Remark 4.12(ii).

(iii) Determine a set $\mathcal{R}$ with Property 2, see Definition 3.5. Fix a bijection $\varphi \colon \mathcal{R} \to \{0, \ldots, |\mathcal{R}| - 1\}$. Thus the map

$$\psi \colon U_{\mathrm{iso}}(\mathcal{R}) \hookrightarrow \mathbb{R}^{d|\mathcal{R}|}, \quad u \mapsto (u(\varphi^{-1}(0)), \ldots, u(\varphi^{-1}(|\mathcal{R}| - 1)))^{\mathsf{T}},$$

which maps a function to a column vector, is an embedding, where $U_{\mathrm{iso}}(\mathcal{R})$ is defined in Definition 3.1. By Proposition 3.28 and the Gram-Schmidt process, we can determine an orthonormal basis $\{b_1, \ldots, b_n\}$ of $\psi(U_{\mathrm{iso}}(\mathcal{R}))$, where $n = \dim(U_{\mathrm{iso}}(\mathcal{R}))$. Let B be the $d|\mathcal{R}|$-by-n matrix $(b_1, \ldots, b_n)$. The matrix $I_{d|\mathcal{R}|} - BB^{\mathsf{T}}$ is the orthogonal projection matrix with kernel $\psi(U_{\mathrm{iso}}(\mathcal{R}))$. Now we can determine the function $g_{\mathcal{R}}$, i.e. the matrix $g_{\mathcal{R}}(g)$ for all $g \in \mathcal{R}$, see Definition 4.43 and Remark 4.44.

(iv) Determine a representation set R of $\widehat{\mathcal{TF}}/\sim$, e.g. with Lemma 2.36, where $\sim$ is the equivalence relation defined in Definition 2.32. For all $\rho \in R$ determine the space group $\mathcal{G}_\rho$, see Definition 2.38 and Definition 4.53, with, e.g., Proposition 2.39, and determine a representation set (or a fundamental domain, see Remark 4.55(i)) K_ρ of $\mathbb{R}^{d_2}/\mathcal{G}_\rho$.

(v) Fix a complete set of representatives of the cosets of $\mathcal{TF}$ in $\mathcal{G}$. Thus the induced representation $\operatorname{Ind}(\chi_k \rho)$ is well-defined for all $\rho \in R$ and $k \in K_\rho$, see Definition 2.29 and Definition 2.23. For all $\rho \in R$ and $k \in K_\rho$ the matrices $\widehat{f_V}(\operatorname{Ind}(\chi_k \rho))$ and $\widehat{g_{\mathcal{R}}}(\operatorname{Ind}(\chi_k \rho))$ can be computed with Definition 2.59. For all $\rho \in R$ and all but finitely many $k \in K_\rho$, the matrix $\widehat{g_{\mathcal{R}}}(\operatorname{Ind}(\chi_k \rho))$ has full rank and thus the real number $\lambda_{\min}(\widehat{f_V}(\operatorname{Ind}(\chi_k \rho)), \widehat{g_{\mathcal{R}}}(\operatorname{Ind}(\chi_k \rho)))$ can easily be computed, see Definition 4.49 and Remark 4.50(ii). Due to the upper semicontinuity, see Lemma 4.52, by Theorem 4.54 we can compute the extended real number λ_{a}.

(vi) The triple $(\mathcal{G}, x_0, V)$ is stable with respect to $\|\cdot\|_{\mathcal{R}}$ if and only if $\chi_{\mathcal{G}} x_0$ is a critical point of E and $\lambda_{\mathrm{a}} > 0$, see Definition 4.8.

In the following two examples, we investigate the stability of a triple $(\mathcal{G}_i, x_i, V_i)$ for all $i \in I$, where I is a suitable index set. The figures are generated with the programming language Python, see https://github.com/Toymodel-Nanotube/ for the source code.

Example 4.57. A suitable toy model for the investigation of stability is an atom chain.

Let $a > 0$ be the scale factor, $t = t_a = (I_2, ae_2) \in \mathrm{E}(2)$ and $\mathcal{G} = \mathcal{G}_a = \langle t \rangle < \mathrm{E}(2)$ analogously to Definition 2.63 and Definition 2.63. We define the interaction potential $V = V_a$, see Definition 4.1 and Remark 4.2(iv), by

$$V_a(y) = v_1(\|y(t_a)\|) + v_2(\|y(t_a^2)\|),$$

where

$$v_1 \colon (0, \infty) \to \mathbb{R}, \quad r \mapsto r^{-12} - r^{-6}$$

is the Lennard-Jones potential and

$$v_2 \colon (0, \infty) \to \mathbb{R}, \quad r \mapsto 8r^{-6}.$$

Let $x_0 = 0_2$. By Lemma 4.6 for all $a > 0$ we have

$$E(\chi_{\mathcal{G}} x_0) = V(y_0) = a^{-12} - \frac{7}{8} a^{-6},$$

where $E = E_a$ is the configurational energy and $y_0 = y_{0,a} = (g \cdot x_0 - x_0)_{g \in \mathcal{G}_a}$. We define

$$a^* := \arg\min_{a \in (0, \infty)} E(\chi_{\mathcal{G}} x_0) = \sqrt[6]{\frac{16}{7}} \approx 1.1477.$$

Thus the structure $\mathcal{G} \cdot x_0$ is stretched (resp. compressed) if $a > a^*$ (resp. $a < a^*$). Now we investigate its stability numerically with Algorithm 4.56.

(i) By Corollary 4.17 the function $\chi_{\mathcal{G}} x_0$ is a critical point of E for all $a > 0$.

(ii) We have

$$\partial_g \partial_h V(y_0) = \begin{cases} 6a^{-8} \begin{pmatrix} -2a^{-6} + 1 & 0 \\ 0 & 26a^{-6} - 7 \end{pmatrix} & \text{if } g = h = t \\[2ex] 2^{-4} 3a^{-8} \begin{pmatrix} -1 & 0 \\ 0 & 7 \end{pmatrix} & \text{if } g = h = t^2 \\[2ex] 0_{2,2} & \text{else.} \end{cases}$$

We have $(\{id\} \cup \operatorname{supp} V)^{-1}(\{id\} \cup \operatorname{supp} V) = \{t^{-2}, \ldots, t^2\}$ and

$$f_V(g) = \begin{cases} a^{-8} \begin{pmatrix} -24a^{-6} + 93/8 & 0 \\ 0 & 312a^{-6} - 651/8 \end{pmatrix} & \text{if } g = id \\[12pt] 6a^{-8} \begin{pmatrix} 2a^{-6} - 1 & 0 \\ 0 & -26a^{-6} + 7 \end{pmatrix} & \text{if } g \in \{t^{-1}, t\} \\[12pt] 2^{-4} 3a^{-8} \begin{pmatrix} 1 & 0 \\ 0 & -7 \end{pmatrix} & \text{if } g \in \{t^{-2}, t^2\} \\[12pt] 0_{2,2} & \text{else.} \end{cases}$$

(iii) Since the set $\{id, t\}$ has Property 1 and $\{t\}$ generates $\mathcal{G}$, the set $\mathcal{R} = \{id, t, t^2\}$ has Property 2. We define the functions

$$b_i \colon \mathcal{R} \to \mathbb{R}^2, \quad g \mapsto e_i \qquad\qquad \text{for all } i \in \{1, 2\}$$

and

$$b_3 \colon \mathcal{R} \to \mathbb{R}^2, \quad g \mapsto \begin{pmatrix} 0 & -1 \\ 1 & 0 \end{pmatrix} (g \cdot x_0 - x_0).$$

By Proposition 3.28 the sets $\{b_1, b_2, b_3\}$ and $\{b_1, b_2\}$ are bases of $U_{\mathrm{iso}}(\mathcal{R})$ and $U_{\mathrm{iso},0,0}(\mathcal{R})$, respectively. We define the bijection $\varphi \colon \mathcal{R} \to \{0, 1, 2\}$ by $t^n \mapsto n$ for all $n \in \{0, 1, 2\}$. Let ψ be the embedding

$$U_{\mathrm{iso}}(\mathcal{R}) \hookrightarrow \mathbb{R}^6, \quad u \mapsto (u(\varphi^{-1}(0)), \ldots, u(\varphi^{-1}(2))).$$

A computation shows that the orthogonal projection matrices of $\mathbb{R}^6$ with kernels $\psi(U_{\mathrm{iso}}(\mathcal{R}))$ and $\psi(U_{\mathrm{iso},0,0}(\mathcal{R}))$ are

$$\frac{1}{6} \begin{pmatrix} 1 & 0 & -2 & 0 & 1 & 0 \\ 0 & 4 & 0 & -2 & 0 & -2 \\ -2 & 0 & 4 & 0 & -2 & 0 \\ 0 & -2 & 0 & 4 & 0 & -2 \\ 1 & 0 & -2 & 0 & 1 & 0 \\ 0 & -2 & 0 & -2 & 0 & 4 \end{pmatrix}$$

and

$$\frac{1}{3} \begin{pmatrix} 2 & 0 & -1 & 0 & -1 & 0 \\ 0 & 2 & 0 & -1 & 0 & -1 \\ -1 & 0 & 2 & 0 & -1 & 0 \\ 0 & -1 & 0 & 2 & 0 & -1 \\ -1 & 0 & -1 & 0 & 2 & 0 \\ 0 & -1 & 0 & -1 & 0 & 2 \end{pmatrix},$$

respectively. Thus the functions $g_{\mathcal{R}}$ and $g_{\mathcal{R},0,0}$ of Definition 4.43 are given by

$$\operatorname{supp} g_{\mathcal{R}} = \mathcal{R}, \qquad g_{\mathcal{R}}(id) = \frac{1}{6}\begin{pmatrix} 1 & 0 \\ 0 & 4 \\ -2 & 0 \\ 0 & -2 \\ 1 & 0 \\ 0 & -2 \end{pmatrix},$$

$$g_{\mathcal{R}}(t) = \frac{1}{6}\begin{pmatrix} -2 & 0 \\ 0 & -2 \\ 4 & 0 \\ 0 & 4 \\ -2 & 0 \\ 0 & -2 \end{pmatrix}, \text{ and } \quad g_{\mathcal{R}}(t^2) = \frac{1}{6}\begin{pmatrix} 1 & 0 \\ 0 & -2 \\ -2 & 0 \\ 0 & -2 \\ 1 & 0 \\ 0 & 4 \end{pmatrix},$$

and

$$\operatorname{supp} g_{\mathcal{R},0,0} = \mathcal{R}, \qquad g_{\mathcal{R},0,0}(id) = \frac{1}{3}\begin{pmatrix} 2 & 0 \\ 0 & 2 \\ -1 & 0 \\ 0 & -1 \\ -1 & 0 \\ 0 & -1 \end{pmatrix},$$

$$g_{\mathcal{R},0,0}(t) = \frac{1}{3}\begin{pmatrix} -1 & 0 \\ 0 & -1 \\ 2 & 0 \\ 0 & 2 \\ -1 & 0 \\ 0 & -1 \end{pmatrix}, \text{ and } g_{\mathcal{R},0,0}(t^2) = \frac{1}{3}\begin{pmatrix} -1 & 0 \\ 0 & -1 \\ -1 & 0 \\ 0 & -1 \\ 2 & 0 \\ 0 & 2 \end{pmatrix},$$

respectively.

(iv) We have $\mathcal{G} = \mathcal{TF} = \langle t \rangle$, $M_0 = \mathbb{N}$ and $\{id\}$ is a representation set of $\widehat{\mathcal{TF}}/\sim$ by Lemma 2.36(i). Recall Definition 2.28. We have $\mathcal{S} = \langle (I_1, a) \rangle$, $L_{\mathcal{S}} = \langle a \rangle$ and $L_{\mathcal{S}}^* = \langle a^{-1} \rangle$. By Proposition 2.39 we have $\{k \in \mathbb{R} \mid (I_1, k) \in \mathcal{G}_{id}\} = \langle a^{-1} \rangle$ and thus $\mathcal{G}_{id} = \langle (I_1, a^{-1}) \rangle$. The interval $K_{id} = [0, a^{-1})$ is a representation set of $\mathbb{R}/\mathcal{G}_{id}$.

(v) For all $k \in K_{id}$ we have $\operatorname{Ind}_{\mathcal{TF}}^{\mathcal{G}} \chi_k = \chi_k$. We have

$$\{k \in K_{id} \mid \widehat{g_{\mathcal{R}}}(\chi_k) \text{ has full rank}\} = K_{id} \setminus \{0\}$$

and
$$\{k \in K_{id} \,|\, \widehat{g_{\mathcal{R},0,0}}(\chi_k) \text{ has full rank}\} = K_{id} \setminus \{0\}.$$

For all $k \in K_{id} \setminus \{0\}$ we can compute $\lambda_{\min}(\widehat{f_V}(\chi_k), \widehat{g_{\mathcal{R}}}(\chi_k))$ and $\lambda_{\min}(\widehat{f_V}(\chi_k), \widehat{g_{\mathcal{R},0,0}}(\chi_k))$. In particular we can compute $\lambda_a = \lambda_a(a)$ and $\lambda_{a,0,0} = \lambda_{a,0,0}(a)$ numerically, see Figure 4.1.

(vi) In the compressed case $a \in (0, a^*)$ we have $\lambda_a = -\infty$ and $\lambda_{a,0,0} \in (-\infty, 0)$ and thus $(\mathcal{G}, x_0, V)$ is not stable with respect to both $\|\cdot\|_{\mathcal{R}}$ and $\|\cdot\|_{\mathcal{R},0,0}$. Now we investigate the stretched case, i.e. $a > a^*$. Let $a^{**} = \sqrt[6]{26/7} \approx 1.244455$. For all $a \in (a^*, a^{**})$ we have $\lambda_a > 0$ and $\lambda_{a,0,0} > 0$ and thus $(\mathcal{G}, x_0, V)$ is stable with respect to both $\|\cdot\|_{\mathcal{R}}$ and $\|\cdot\|_{\mathcal{R},0,0}$. For all $a > a^{**}$ we have $E''(\chi_{\mathcal{G}}x_0)(u, u) < 0$, where $u = e_2 \chi_{\{t^n \,|\, n \in 2\mathbb{Z}\}}$. In particular we have $\lambda_a < 0$ and $\lambda_{a,0,0} < 0$ and thus $(\mathcal{G}, x_0, V)$ is not stable with respect to both $\|\cdot\|_{\mathcal{R}}$ and $\|\cdot\|_{\mathcal{R},0,0}$ for all $a > a^{**}$.

Notice that in the stretched case $a \in (a^*, a^{**})$, the appropriate seminorm for the stability is $\|\cdot\|_{\mathcal{R},0,0}$. For the equilibrium case $a \approx a^*$, the weaker seminorm $\|\cdot\|_{\mathcal{R}}$ is appropriate since $\lim_{a \to a^*} \lambda_a = 0$ and $\lim_{a \searrow a^*} \lambda_{a,0,0} > 0$.

Example 4.58. There exists a huge literature on the stability of (n, m) nanotubes as zigzag or armchair nanotubes, see, e.g., [30]. Each (n, m) nanotube is the orbit of some point in $\mathbb{R}^3$ under the action of a discrete subgroup of E(3). Thus its stability can be checked with Algorithm 4.56. In this example we investigate the stability of a $(5, 1)$ nanotube, see Figure 4.2.

For all scale factors $a > 0$ and angles $\alpha \in (0, \pi)$ we define: Let $R(\alpha) \in O(2)$ be the rotation matrix as in (2.1), $t = t_{a,\alpha} = (R(\alpha) \oplus I_1, ae_3) \in E(3)$, $p = (I_1 \oplus (-I_2), 0_3) \in E(3)$ and $\mathcal{G} = \mathcal{G}_{a,\alpha}$ be the discrete group $\langle t, p \rangle < E(3)$, i.e. $\mathcal{G} = \{t^m p^q \,|\, m \in \mathbb{Z}, q \in \{0, 1\}\}$. For all $x \in \mathbb{R}^3$ we have $\mathcal{G} \cdot x \subset C_x$, where C_x is the cylinder $\{y \in \mathbb{R}^3 \,|\, y_1^2 + y_2^2 = x_1^2 + x_2^2\}$.

Let $\mathcal{N} = \mathcal{N}_{a,\alpha} = \{tp, t^6 p, t^7 p\}$. Let $U_{a,\alpha} \subset \mathbb{R}^3$ be the set of all points $x \in \mathbb{R}^3$ for which the map $\mathcal{G} \to \mathbb{R}^3$, $g \mapsto g \cdot x$ is injective and the three nearest neighbors of x in $\mathcal{G} \cdot x$ are the points $\mathcal{N} \cdot x$, i.e.

$$\sup\{\|g \cdot x - x\| \,\big|\, g \in \mathcal{N}\} < \inf\{\|g \cdot x - x\| \,\big|\, g \in \mathcal{G} \setminus (\mathcal{N} \cup \{id\})\}.$$

Let
$$W := \{(a, \alpha, x) \,|\, a > 0, \alpha \in (0, \pi), x \in U_{a,\alpha}\}.$$

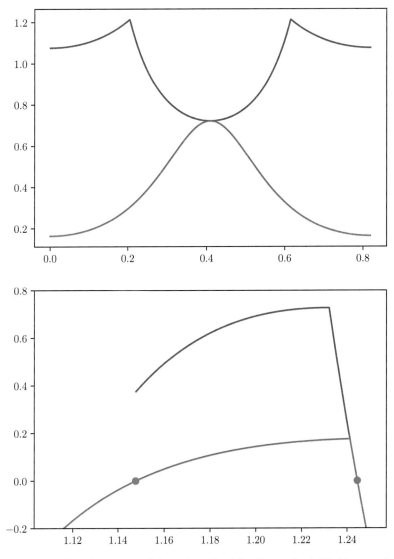

Figure 4.1.: For the toy model as described in Example 4.57, the graphs of $\lambda_{\min}(\widehat{f_V}(\chi_k), \widehat{g_{\mathcal{R}}}(\chi_k))$ (blue) and $\lambda_{\min}(\widehat{f_V}(\chi_k), \widehat{g_{\mathcal{R},0,0}}(\chi_k))$ (orange) dependent on $k \in K_{id} \setminus \{0\}$ are plotted on the top plot for the choice $a = 1.22$. The points $(a^*, 0)$ and $(a^{**}, 0)$ and the graphs of λ_{a} (blue) and $\lambda_{\mathrm{a},0,0}$ (orange) dependent on the scale factor a are plotted on the bottom plot.

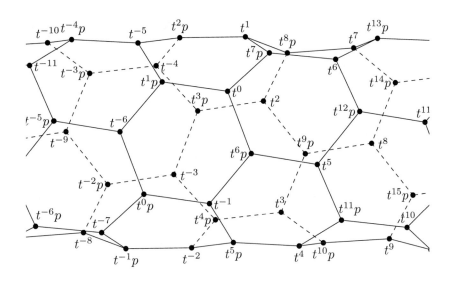

Figure 4.2.: As described in Example 4.57, the orbit of the point x_{a_0}
under the action of the group $\mathcal{G}_{a_0,\alpha_0}$ is a $(5,1)$ nanotube. We
have a natural bijection between the group elements and the
atoms.

Analogously to [30] we define the interaction potential $V = V_{a,\alpha}$, see Definition 4.1 and Remark 4.2(iv), by

$$V(y) = \frac{1}{2} \sum_{g \in \mathcal{N}} v_1(\|y(g)\|) + \frac{1}{2} \sum_{g,h \in \mathcal{N}} v_2(y(g), y(h)),$$

where

$$v_1 \colon (0,\infty) \to \mathbb{R}, r \mapsto (r-1)^2$$

is a *two-body potential* and

$$v_2 \colon \{(x,y) \mid x,y \in \mathbb{R}^3 \setminus \{0\}\} \to \mathbb{R}, (x,y) \mapsto \left(\frac{\langle x,y \rangle}{\|x\|\|y\|} + \frac{1}{2} \right)^2$$

is a *three-body potential*. Thus the *bonded* points of $\mathcal{G} \cdot x$ tend to have distance 1 and the *bond angles* tend to form $2\pi/3$ angles. By Lemma 4.6 for all $(a, \alpha, x) \in W$ we have $E(\chi_{\mathcal{G}} x) = V(y_0)$, where $E = E_{a,\alpha}$ is the configurational energy and $y_0 = y_{0,a,\alpha,x} = (g \cdot x - x)_{g \in \mathcal{G}_{a,\alpha}}$.
First we consider the $(5, 1)$ nanotube. We define

$$\alpha_0 := 11\pi/31 \approx 1.115$$

and

$$x_a := a(r\cos(\beta), r\sin(\beta), 7/3) \in \mathbb{R}^3 \qquad \text{for all } a > 0,$$

where $r = 31/(\pi\sqrt{3})$ and $\beta = 5\pi/31$. With the formulas in [22] it follows that for all $(a, \alpha, x) \in W$ the set $\mathcal{G} \cdot x$ is a so called $(5, 1)$ nanotube if and only if $\alpha = \alpha_0$ and $x = x_a$. The bond length of the unrolled $(5, 1)$ nanotube $\mathcal{G}_{a,\alpha_0} \cdot x_a$, i.e. the distance of two neighbored points of $\mathcal{G}_{a,\alpha_0} \cdot x_a$ with respect to the induced metric of the manifold C_x, is equal to 1 if and only if $a = a_0$, where

$$a_0 := 3/(2\sqrt{31}) \approx 0.269.$$

Now we investigate numerically with Algorithm 4.56 the stability of the $(5, 1)$ nanotube, more precisely of $(\mathcal{G}_{a,\alpha_0}, x_a, V_{a,\alpha_0})$.

(i) For all $a > 0$ we have $e_{V_{a,\alpha_0}} \neq 0$, see Figure 4.3, and thus $\chi_{\mathcal{G}_{a,\alpha_0}} x_a$ is not a critical point of E_{a,α_0}. Thus we can proceed with (vi).

(vi) By (i) for all $a > 0$ the triple $(\mathcal{G}_{a,\alpha_0}, x_a, V_{a,\alpha_0})$ is not stable with respect to both $\|\cdot\|_{\mathcal{R}}$ and $\|\cdot\|_{\mathcal{R},0,0}$.

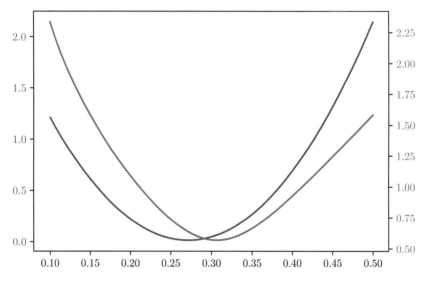

Figure 4.3.: For the $(5, 1)$ nanotube as described in Example 4.58, the graphs of the energy $E(\chi_{g_{a,\alpha_0}} x_a)$ and the norm of $e_{V_{a,\alpha_0}}$ dependent on the scale factor a are plotted in blue and orange, respectively. For all a, we have $e_{V_{a,\alpha_0}} \neq 0$ and thus the $(5, 1)$ nanotube is not stable.

We define

$$(a^*, \alpha^*, x^*) := \underset{(a,\alpha,x) \in W}{\arg\min} E_{a,\alpha}(\chi_{\mathcal{G}_{a,\alpha}} x)$$

$$\approx (0.263, 1.117, (1.388, 0.776, 0.626))$$

and

$$x_a^* := \underset{x \in U_{a,\alpha^*}}{\arg\min} E(\chi_{\mathcal{G}} x) \qquad \text{for all } a \approx a^*.$$

In particular we have $x^* = x_{a^*}^*$. We have $(a^*, \alpha^*, x^*) \approx (a_0, \alpha_0, x_{a_0})$ and thus the nanotube $\mathcal{G}_{a^*,\alpha^*} \cdot x^*$ is approximately equal to the $(5,1)$ nanotube $\mathcal{G}_{a_0,\alpha_0} \cdot x_{a_0}$. Now for all $a \approx a^*$ we check the stability of $(\mathcal{G}_{a,\alpha^*}, x_a^*, V_{a,\alpha^*})$ numerically with Algorithm 4.56.

(i) For all $a \approx a^*$ the function $\chi_{\mathcal{G}} x_a^*$ is a critical point of E by Remark 4.15(ii) and Corollary 4.16.

(ii) We have

$$\operatorname{supp} V = \{tp, t^6 p, t^7 p\}$$

and

$$\operatorname{supp} f_V = \{t^{-6}, t^{-5}, t^{-1}, id, t, t^5, t^6, tp, t^6 p, t^7 p\}$$

by Remark 4.15(ii). The first and second derivative of V can be computed, e.g., with the Python library SymPy and f_V can be computed numerically by Definition 4.11.

(iii) Since $\{t^{-1}, id, t, p\}$ has Property 1 and $\{t, p\}$ generates $\mathcal{G}$, by Definition 3.5 the set

$$\mathcal{R} = \mathcal{R}_a := \{t^{-1}, id, t, t^2, t^{-1} p, p, tp\}$$

has Property 2. We define the bijection φ between $\mathcal{R}$ and $\{0, \dots, 6\}$ by $\varphi(t^m) = m + 1$ for all $m \in \{-1, 0, 1, 2\}$ and $\varphi(t^m p) = m + 5$ for all $m \in \{-1, 0, 1\}$. For all $a \approx a^*$ we define the functions

$$b_i = b_{i,a} \colon \mathcal{R} \to \mathbb{R}^3, \quad g \mapsto L(g)^\mathsf{T} e_i \qquad \text{for all } i \in \{1, 2, 3\}$$

and

$$b_i = b_{i,a} \colon \mathcal{R} \to \mathbb{R}^3, \quad g \mapsto L(g)^\mathsf{T} A_i (g \cdot x_a^* - x_a^*) \quad \text{for all } i \in \{4, 5, 6\},$$

where

$$A_4 = \begin{pmatrix} 0 & -1 & 0 \\ 1 & 0 & 0 \\ 0 & 0 & 0 \end{pmatrix}, A_5 = \begin{pmatrix} 0 & 0 & -1 \\ 0 & 0 & 0 \\ 1 & 0 & 0 \end{pmatrix} \text{ and } A_6 = \begin{pmatrix} 0 & 0 & 0 \\ 0 & 0 & -1 \\ 0 & 1 & 0 \end{pmatrix}.$$

By Proposition 3.28 the sets $\{b_1, \ldots, b_6\}$ and $\{b_1, \ldots, b_4\}$ are bases of $U_{\mathrm{iso}}(\mathcal{R})$ and $U_{\mathrm{iso},0,0}(\mathcal{R})$, respectively. With, e.g., the Gram-Schmidt process we can determine functions $b_1', \ldots, b_6' \colon \mathcal{R} \to \mathbb{R}^3$ such that $\{b_1', \ldots, b_6'\}$ and $\{b_1', \ldots, b_4'\}$ are orthonormal bases of $U_{\mathrm{iso}}(\mathcal{R})$ and $U_{\mathrm{iso},0,0}(\mathcal{R})$, respectively. A bijection between $\{u \colon \mathcal{R} \to \mathbb{C}^3\}$ and $\mathbb{C}^{21}$ is given by $u \mapsto (u(\varphi^{-1}(0)), \ldots, u(\varphi^{-1}(6)))$. Let $B = (b_1', \ldots, b_6') \in \mathbb{R}^{21 \times 6}$ and $B_0 = (b_1', \ldots, b_4') \in \mathbb{R}^{21 \times 4}$. The matrices $P = I_{21} - BB^{\mathsf{T}}$ and $P_0 = I_{21} - B_0 B_0^{\mathsf{T}}$ are orthogonal projection matrices with kernels $U_{\mathrm{iso}}(\mathcal{R})$ and $U_{\mathrm{iso},0,0}(\mathcal{R})$, respectively. Let $p_0, \ldots, p_6, p_{0,0}, \ldots, p_{0,6} \in \mathbb{R}^{21 \times 3}$ such that $P = (p_0, \ldots, p_6)$ and $P_0 = (p_{0,0}, \ldots, p_{0,6})$. For the functions $g_{\mathcal{R}}$ and $g_{\mathcal{R},0,0}$ of Definition 4.43 we have

$$\operatorname{supp} g_{\mathcal{R}} = \operatorname{supp} g_{\mathcal{R},0,0} = \mathcal{R},$$
$$g_{\mathcal{R}}(g) = p_{\varphi(g)} \qquad\qquad \text{for all } g \in \mathcal{R}$$

and

$$g_{\mathcal{R},0,0}(g) = p_{0,\varphi(g)} \qquad\qquad \text{for all } g \in \mathcal{R}.$$

(iv) We have $\mathcal{TF} = \mathcal{T} = \langle t \rangle$, $M_0 = \mathbb{N}$ and $\{id\}$ is a representation set of $\widehat{\mathcal{TF}}/\sim$ by Lemma 2.36(i). Recall Definition 2.28. We have $L_{\mathcal{S}} = \langle a \rangle$ and $L_{\mathcal{S}}^* = \langle a^{-1} \rangle$. By Proposition 2.39 we have $\{k \in \mathbb{R} \mid (I_1, k) \in \mathcal{G}_{id}\} = \langle a^{-1} \rangle$ and thus $\mathcal{G}_{id} = \{((-I_1)^q, ma^{-1}) \mid m \in \mathbb{Z}, q \in \{0,1\}\}$. The interval $K_{id} = [0, 1/(2a))$ is a representation set of $\mathbb{R}/\mathcal{G}_{id}$.

(v) The set $\{id, p\}$ is a complete set of representatives of the cosets of $\mathcal{TF}$ in $\mathcal{G}$. For all $k \in K_{id}$ and $g \in \mathcal{G}$ we have

$$\operatorname{Ind}_{\mathcal{TF}}^{\mathcal{G}} \chi_k(g) = \begin{cases} \begin{pmatrix} \chi_k(g) & 0 \\ 0 & \chi_k(p^{-1}gp) \end{pmatrix} & \text{if } g \in \mathcal{TF} \\[2ex] \begin{pmatrix} 0 & \chi_k(gp) \\ \chi_k(p^{-1}g) & 0 \end{pmatrix} & \text{else.} \end{cases}$$

Now for all $k \in K_{id}$, it is easy to compute the complex 6-by-6 matrices $\widehat{f_V}(\operatorname{Ind} \chi_k)$, $\widehat{g_{\mathcal{R}}}(\operatorname{Ind} \chi_k)$ and $\widehat{g_{\mathcal{R},0,0}}(\operatorname{Ind} \chi_k)$. We have

$$\{k \in K_{id} \mid \widehat{g_{\mathcal{R}}}(\operatorname{Ind} \chi_k) \text{ has full rank}\} = K_{id} \setminus \{0, \alpha^*/(2\pi a)\}$$

and

$$\{k \in K_{id} \mid \widehat{g_{\mathcal{R},0,0}}(\operatorname{Ind} \chi_k) \text{ has full rank}\} = K_{id} \setminus \{0, \alpha^*/(2\pi a)\}.$$

For all $k \in K_{id} \setminus \{0, \alpha^*/(2\pi a)\}$ we can compute $\lambda_{\min}(\widehat{f_V}(\operatorname{Ind}\chi_k), \widehat{g_R}(\operatorname{Ind}\chi_k))$ and $\lambda_{\min}(\widehat{f_V}(\operatorname{Ind}\chi_k), \widehat{g_{R,0,0}}(\operatorname{Ind}\chi_k))$. In particular we can compute $\lambda_a(a, \alpha^*)$ and $\lambda_{a,0,0}(a, \alpha^*)$ numerically, see Figure 4.4.

(vi) In the stretched case $a > a^*$, we have both $\lambda_a(a, \alpha^*) > 0$ and $\lambda_{a,0,0}(a, \alpha^*) > 0$ and thus $(\mathcal{G}_{a,\alpha^*}, x_{a,\alpha^*}, V_{a,\alpha^*})$ is stable with respect to both $\|\cdot\|_R$ and $\|\cdot\|_{R,0,0}$. In the compressed case $a \in (0, a^*)$ we have $\lambda_a(a, \alpha^*) = -\infty$ and $\lambda_{a,0,0}(a, \alpha^*) < 0$ and thus $(\mathcal{G}_{a,\alpha^*}, x_{a,\alpha^*}, V_{a,\alpha^*})$ is not stable with respect to both $\|\cdot\|_R$ and $\|\cdot\|_{R,0,0}$.

Notice that in the stretched case $a > a^*$, the appropriate seminorm for the stability is $\|\cdot\|_{R,0,0}$. For the equilibrium case $a \approx a^*$, the weaker seminorm $\|\cdot\|_R$ is appropriate since $\lim_{a \searrow a^*} \lambda_{a,0,0}(a, \alpha^*) = 0$ and $\lim_{a \searrow a^*} \lambda_a(a, \alpha^*) > 0$.

For all $a \approx a^*$ and $\alpha \approx \alpha^*$ we can compute $\lambda_a(a, \alpha)$ and $\lambda_{a,0,0}(a, \alpha)$ analogously. For $\alpha \approx \alpha^*$ the graphs of $\lambda_a(\cdot, \alpha)$ and $\lambda_{a,0,0}(\cdot, \alpha)$ are similar to the graphs of $\lambda_a(\cdot, \alpha^*)$ and $\lambda_{a,0,0}(\cdot, \alpha^*)$. As an example, we consider

$$\alpha_a := \arg\min_{\alpha \in (0,\pi)} E(\chi_{\mathcal{G}} x_{a,\alpha}) \qquad \text{for all } a \approx \alpha^*,$$

see Figure 4.5. In Figure 4.5 the graphs of the functions

$$a \mapsto \text{Relative difference}\big(\lambda_a(a, \alpha^*), \lambda_a(a, \alpha_a)\big)$$

and

$$a \mapsto \text{Relative difference}\big(\lambda_{a,0,0}(a, \alpha^*), \lambda_{a,0,0}(a, \alpha_a)\big)$$

are plotted, where

$$\text{Relative difference}(x, y) := |x - y|/\max\{|x|, |y|\} \qquad \text{for all } x, y \in \mathbb{R}.$$

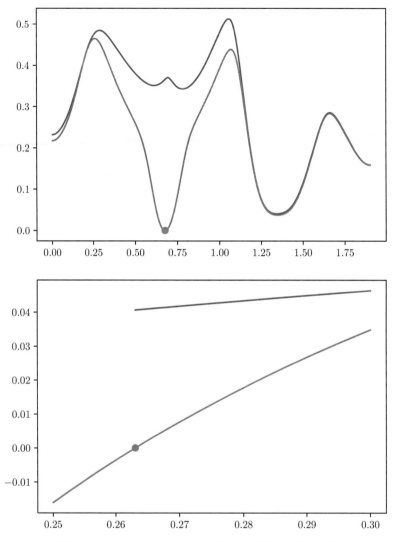

Figure 4.4.: For the nanotube as described in Example 4.58, the point $(\alpha^*/(2\pi a^*), 0)$ and the graphs of $\lambda_{\min}(\widehat{f_V}(\chi_k), \widehat{g_\mathcal{R}}(\chi_k))$ (blue) and $\lambda_{\min}(\widehat{f_V}(\chi_k), \widehat{g_{\mathcal{R},0,0}}(\chi_k))$ (orange) dependent on $k \in K_{id} \setminus \{0, \alpha^*/(2\pi a^*)\}$ are plotted on the top plot for the choice $a = a^*$. The point $(a^*, 0)$ and and the graphs of λ_{a} (blue) and $\lambda_{\mathrm{a},0,0}$ (orange) dependent on the scale factor are plotted on the bottom plot.

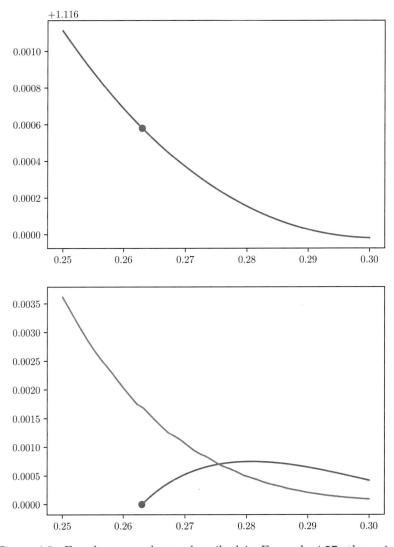

Figure 4.5.: For the nanotube as described in Example 4.57, the point $(a^*, \alpha(a^*))$ and the graph of the angle $\alpha(a)$ dependent on the scale factor a are plotted on the top plot. The point $(a^*, 0)$ and the graphs of $\text{Relative difference}\big(\lambda_a(a, \alpha^*), \lambda_a(a, \alpha_a)\big)$ (blue) and $\text{Relative difference}\big(\lambda_{a,0,0}(a, \alpha^*), \lambda_{a,0,0}(a, \alpha_a)\big)$ (orange) dependent on the scale factor a are plotted on the bottom plot.

A. The configurational energy restricted to $U_{\mathrm{iso},0,0} \cap U_{\mathrm{per}}$

In the following we prove Remark 4.30(iii), see Proposition A.3. Proposition A.3 is similar to Proposition 4.29.

Lemma A.1. *Suppose that V is weakly* sequentially continuous. Then for all functions $y\colon \mathcal{G} \setminus \{id\} \to \mathbb{R}^d$ and constants $C, c > 0$ there exists a finite set $\mathcal{A} \subset \mathcal{G} \setminus \{id\}$ such that*

$$|V(y + z) - V(y)| < c$$

for all $z \in L^\infty(\mathcal{G} \setminus \{id\}, \mathbb{R}^d)$ with $\|z\|_\infty \leq C$ and $z(g) = 0$ for all $g \in \mathcal{A}$.

Proof. This is clear since V is weakly* sequentially continuous and by Exercise 2.51b) in [37]. $\qquad\square$

Remark A.2. A sequence $(y_n)_{n \in \mathbb{N}}$ in $L^\infty(\mathcal{G} \setminus \{id\}, \mathbb{R}^d)$ converges to $y \in L^\infty(\mathcal{G} \setminus \{id\}, \mathbb{R}^d)$ with respect to the weak* topology if and only if the sequence $(y_n)_{n \in \mathbb{N}}$ is bounded and $(y_n)_{n \in \mathbb{N}}$ converges componentwise to y, i.e. $\lim_{n \to \infty} y_n(g) = y(g)$ for all $g \in \mathcal{G} \setminus \{id\}$, see Exercise 2.51 in [37].

Proposition A.3. *Suppose that V is weakly* sequentially continuous, $E'(\chi_{\mathcal{G}} x_0) = 0$ and let $u \in U_{\mathrm{iso},0,0} \cap U_{\mathrm{per}}$. Then it holds $E''(\chi_{\mathcal{G}} x_0)(u, u) = 0$ and $\frac{d^3}{d\tau^3} E(\chi_{\mathcal{G}} x_0 + \tau u)\big|_{\tau=0} = 0$.*

Proof. Suppose that V is weakly* sequentially continuous and $E'(\chi_{\mathcal{G}} x_0) = 0$. Thus for the monotonically increasing function

$$\begin{aligned} r\colon [0, \infty) &\to [0, \infty) \\ t &\mapsto \sup\left\{ |E(\chi_{\mathcal{G}} x_0 + u) - E(\chi_{\mathcal{G}} x_0)| \,\big|\, u \in B_t(0) \right\} \end{aligned}$$

it holds

$$\lim_{t \searrow 0} \frac{r(t)}{t^2} = \sup\{E''(\chi_{\mathcal{G}} x_0)(u, u) \,|\, u \in U_{\mathrm{per}}\} < \infty, \tag{A.1}$$

where $B_t(0) = \{u \in U_{\mathrm{per}} \,|\, \|u\|_\infty < t\}$ for all $t > 0$.

Let $u \in U_{\text{iso},0,0} \cap U_{\text{per}}$. There exist some $a \in \mathbb{R}^d$ and $S \in \oplus(\text{Skew}(d_1) \times \{0_{d_2,d_2}\})$ such that

$$L(g)u(g) = a + S(g \cdot x_0 - x_0) \quad \text{for all } g \in \mathcal{G}.$$

Since differentiability implies locally boundedness, there exist some $\delta > 0$ and $C_1 > 0$ such that

$$|V(y_0 + w)| \le C_1 \qquad \text{for all } w \in B_\delta(0),$$

where $B_\delta(0) = \{w \in L^\infty(\mathcal{G} \setminus \{id\}, \mathbb{R}^d) \,|\, \|w\|_\infty < \delta\}$. Let

$$C_2 = 2\|x_0\| \sup\{\|\mathrm{e}^{-\tau S} - I_d + \tau S\mathrm{e}^{-\tau S}\|/\tau^2 \,|\, \tau \in (-1,1)\} \ge 0.$$

By Taylor's theorem we have $C_2 < \infty$. Let $t_0 = \min\{1, \sqrt{\delta/(2C_2)}\} > 0$, where $a/0 := \infty$ for all $a > 0$.
Now we show that

$$|E(\chi_\mathcal{G} x_0 + tu) - E(\chi_\mathcal{G} x_0)| \le r(C_2 t^2) + t^4 \qquad \text{for all } t \in (-t_0, t_0). \quad \text{(A.2)}$$

Let $t \in (-t_0, t_0) \setminus \{0\}$. We define the function $v \colon \mathcal{G} \to \mathbb{R}^d$ by

$$g \cdot v(g) = x_0 + \mathrm{e}^{-tS}(I_d + tS)(g \cdot x_0 - x_0) \qquad \text{for all } g \in \mathcal{G},$$

see also Figure A.1. We have

$$
\begin{aligned}
\|v - \chi_\mathcal{G} x_0\|_\infty &= \sup\{\|v(g) - x_0\| \,|\, g \in \mathcal{G}\} \\
&= \sup\{\|g \cdot v(g) - g \cdot x_0\| \,|\, g \in \mathcal{G}\} \\
&= \sup\{\|(\mathrm{e}^{-tS} - I_d + tS\mathrm{e}^{-tS})(g \cdot x_0 - x_0)\| \,|\, g \in \mathcal{G}\} \\
&= \sup\{\|(\mathrm{e}^{-tS} - I_d + tS\mathrm{e}^{-tS})(L(g)x_0 - x_0)\| \,|\, g \in \mathcal{G}\} \\
&\le 2\|\mathrm{e}^{-tS} - I_d + tS\mathrm{e}^{-tS}\|\|x_0\| \\
&\le C_2 t^2, \qquad\qquad\qquad\qquad\qquad\qquad\qquad\qquad\quad \text{(A.3)}
\end{aligned}
$$

where in the forth step we used that $S\tau(g) = 0$ for all $g \in \mathcal{G}$. In particular, we have $v \in L^\infty(\mathcal{G}, \mathbb{R}^d)$ and

$$\|v - \chi_\mathcal{G} x_0\|_\infty < \frac{\delta}{2}. \qquad\qquad \text{(A.4)}$$

For all $g \in \mathcal{G}$ we define the map

$$
\begin{aligned}
\varphi_g \colon U_{\text{per}} &\to \{w \colon \mathcal{G} \setminus \{id\} \to \mathbb{R}^d\} \\
w &\mapsto (\mathcal{G} \setminus \{id\} \to \mathbb{R}^d, h \mapsto (gh) \cdot w(gh) - g \cdot w(g)).
\end{aligned}
$$

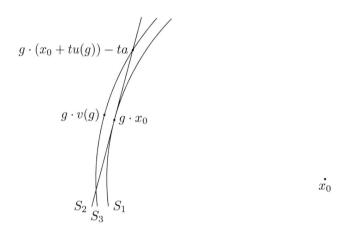

Figure A.1.: In this figure for u, v, a, S and t as in the proof of Lemma A.3 and $g \in \mathcal{G}$, the points x_0, $g \cdot x_0$, $g \cdot (x_0 + u(g)) - a$ and $g \cdot v(g)$ and the sets $S_1 = \{x_0 + A(g \cdot x_0 - x_0) \mid A \in O(d)\}$, $S_2 = \{x_0 + (I_d + \tilde{S})(g \cdot x_0 - x_0) \mid \tilde{S} \in \oplus(\text{Skew}(d_1) \times \{0_{d_2,d_2}\})\}$ and $S_3 = \{x_0 + A(I_d + tS)(g \cdot x_0 - x_0) \mid A \in O(d)\}$ are displaced.

For all $g \in \mathcal{G}$ we have

$$
\begin{aligned}
\varphi_g(\chi_{\mathcal{G}} x_0 + tu) &= \big((gh) \cdot x_0 + tL(gh)u(gh) \\
&\quad - (g \cdot x_0 + tL(g)u(g))\big)_{h \in \mathcal{G} \setminus \{id\}} \\
&= \big((gh) \cdot x_0 + ta + tS((gh) \cdot x_0 - x_0) \\
&\quad - (g \cdot x_0 + ta + tS(g \cdot x_0 - x_0))\big)_{h \in \mathcal{G} \setminus \{id\}} \\
&= \big((I_d + tS)((gh) \cdot x_0 - g \cdot x_0)\big)_{h \in \mathcal{G} \setminus \{id\}} \\
&= \big(e^{tS}((gh) \cdot v(gh) - g \cdot v(g))\big)_{h \in \mathcal{G} \setminus \{id\}} \\
&= e^{tS}\varphi_g(v). \tag{A.5}
\end{aligned}
$$

For all $\mathcal{A} \subset \mathcal{G} \setminus \{id\}$ we denote

$$
B_{\mathcal{A}} := \big\{w \in L^\infty(\mathcal{G} \setminus \{id\}, \mathbb{R}^d) \mid \|w\|_\infty \leq R \text{ and } w(g) = 0 \text{ for all } g \in \mathcal{A}\big\},
$$

where $R = 2(\|x_0\| + t_0 \|u\|_\infty)$. Let $N \in M_0$ such that u is $\mathcal{T}^N$-periodic. Since V is weakly* sequentially continuous, by Lemma A.1 for all $g \in \mathcal{C}_N$

there exists a finite set $\mathcal{A}_g \subset \mathcal{G} \setminus \{id\}$ such that

$$\left| V(\varphi_g(\chi_{\mathcal{G}} x_0 + tu) + w) - V(\varphi_g(\chi_{\mathcal{G}} x_0 + tu)) \right| < \frac{t^4}{2} \text{ for all } w \in B_{\mathcal{A}_g}. \quad (A.6)$$

Let $\mathcal{A} = \bigcup_{g \in \mathcal{C}_N} \mathcal{A}_g$. Equation (A.5), (H1) and (A.6) imply that for all $g \in \mathcal{G}$ we have

$$\sup_{w \in B_{\mathcal{A}}} \left| V(\varphi_g(v) + w) - V(\varphi_g(v)) \right|$$

$$= \sup_{w \in B_{\mathcal{A}}} \left| V(e^{-tS}\varphi_g(\chi_{\mathcal{G}} x_0 + tu) + w) - V(e^{-tS}\varphi_g(\chi_{\mathcal{G}} x_0 + tu)) \right|$$

$$= \sup_{w \in B_{\mathcal{A}}} \left| V(\varphi_g(\chi_{\mathcal{G}} x_0 + tu) + w) - V(\varphi_g(\chi_{\mathcal{G}} x_0 + tu)) \right|$$

$$= \sup_{w \in B_{\mathcal{A}}} \left| V(\varphi_{\tilde{g}}(\chi_{\mathcal{G}} x_0 + tu) + w) - V(\varphi_{\tilde{g}}(\chi_{\mathcal{G}} x_0 + tu)) \right|$$

$$\leq \frac{t^4}{2}, \quad (A.7)$$

where in the third line $\tilde{g} \in \mathcal{G}$ is defined by the condition $\{\tilde{g}\} = g\mathcal{T}^N \cap \mathcal{C}_N$. Let $m \in \mathbb{N}$ such that $M_0 = m\mathbb{N}$. Since $\mathcal{T}^m$ is isomorphic to $\mathbb{Z}^{d_2}$, there exist $t_1, \ldots, t_{d_2} \in \mathcal{T}^m$ such that $\{t_1, \ldots, t_{d_2}\}$ generates $\mathcal{T}^m$. Without loss of generality we assume that $\mathcal{C}_n = \{t_1^{n_1} \ldots t_{d_2}^{n_{d_2}} g \,|\, n_1, \ldots, n_{d_2} \in \{0, \ldots, n/m - 1\}, g \in \mathcal{C}_m\}$ for all $n \in M_0$, see Remark 2.51(ii). There exists some $n' \in \mathbb{N}$ such that

$$\mathcal{C}_m \mathcal{A} \subset \{t_1^{n_1} \ldots t_{d_2}^{n_{d_2}} \,|\, n_1, \ldots, n_{d_2} \in \{-n', \ldots, n'\}\} \mathcal{C}_m.$$

Thus there exists some $N' \in M_0$ such that N divides N' and

$$\frac{|\mathcal{C}_{N'} \setminus \mathcal{D}|}{|\mathcal{C}_{N'}|} < \frac{t^4}{4C_1}, \quad (A.8)$$

where $\mathcal{D} = \{g \in \mathcal{C}_{N'} \,|\, g\mathcal{A} \subset \mathcal{C}_{N'}\}$. We define the $\mathcal{T}^{N'}$-periodic function $\tilde{v} \in U_{\mathrm{per}}$ by

$$\tilde{v}(g) := v(g) \qquad \text{for all } g \in \mathcal{C}_{N'}.$$

It holds

$$|E(\tilde{v}) - E(\chi_{\mathcal{G}} x_0)| \leq r(\|\tilde{v} - \chi_{\mathcal{G}} x_0\|_\infty) \leq r(\|v - \chi_{\mathcal{G}} x_0\|_\infty) \leq r(C_2 t^2), \quad (A.9)$$

where we used (A.3) in the last step. Moreover, we have

$$|E(\chi_{\mathcal{G}} x_0 + tu) - E(\tilde{v})| \leq \frac{1}{|\mathcal{C}_{N'}|} \sum_{g \in \mathcal{C}_{N'}} |V(\varphi_g(\chi_{\mathcal{G}} x_0 + tu)) - V(\varphi_g(\tilde{v}))|$$

$$= \frac{1}{|\mathcal{C}_{N'}|} \sum_{g \in \mathcal{C}_{N'}} |V(e^{tS}\varphi_g(v)) - V(\varphi_g(\tilde{v}))|$$

$$= \frac{1}{|\mathcal{C}_{N'}|} \sum_{g \in \mathcal{C}_{N'}} |V(\varphi_g(v)) - V(\varphi_g(\tilde{v}))|$$

$$\leq \frac{1}{|\mathcal{C}_{N'}|} \sum_{g \in \mathcal{D}} \sup_{w \in B_A} |V(\varphi_g(v)) - V(\varphi_g(v) + w)|$$

$$+ \frac{2}{|\mathcal{C}_{N'}|} \sum_{g \in \mathcal{C}_{N'} \setminus \mathcal{D}} \sup_{w \in B_\delta(0)} |V(\varphi_g(\chi_{\mathcal{G}} x_0) + w)|$$

$$\leq \frac{t^4}{2} + \frac{t^4}{2} = t^4, \tag{A.10}$$

where we used (A.5) in the second step, (H1) in the third step, (A.4) in the forth step and (A.7) and (A.8) in the fifth step. Equation (A.9) and (A.10) imply (A.2).

By (A.2) and (A.1) we have

$$\limsup_{t \to 0} \left| \frac{E(\chi_{\mathcal{G}} x_0 + tu) - E(\chi_{\mathcal{G}} x_0)}{t^3} \right| \leq \limsup_{t \to 0} \frac{r(C_2 t^2)}{t^3} + t = 0$$

and thus, $E''(\chi_{\mathcal{G}} x_0)(u, u) = 0$ and $\frac{d^3}{d\tau^3} E(\chi_{\mathcal{G}} x_0 + \tau u)\big|_{\tau=0} = 0$. $\qquad \square$

B. Representation theory

We need the following propositions in Chapter 2.

In general, the dual space of a locally compact group contains infinite-dimensional representations. In contrast to the rest of the thesis, in the following when we use the term representation, we mean a finite- or infinite-dimensional representation on a Hilbert space.

Proposition B.1 (Proposition 1.35 in [43])**.** *Let ρ be a continuous unitary representation of a locally compact group G on the Hilbert space $\mathcal{H}(\rho)$. Then ρ is irreducible if and only if*

$$\text{commutant of } \rho(G) := \{T \in \mathcal{B}(\mathcal{H}(\rho)) \,|\, T\rho(g) = \rho(g)T \text{ for all } g \in G\}$$
$$= \mathbb{C}I,$$

where $\mathcal{B}(\mathcal{H}(\rho))$ denotes the space of bounded linear operators from $\mathcal{H}(\rho)$ to $\mathcal{H}(\rho)$ and I is the identity operator on $\mathcal{H}(\rho)$.

Proposition B.2 (Proposition 1.71 in [43])**.** *Let N be a closed normal subgroup of a locally compact group G and $q \colon G \to G/N$ be the quotient homomorphism. The map $\rho \mapsto \rho \circ q$ is a homeomorphism of $\widehat{G/N}$ with the closed subset of $\widehat{G}$ consisting of those elements of $\widehat{G}$ which annihilate N.*

C. Seminorms

We need the following definitions and lemma in Chapter 3.

Definition C.1. Given a vector space V over a field $\mathbb{K} \in \{\mathbb{R}, \mathbb{C}\}$, a *seminorm* is a function $p: V \to [0, \infty)$ such that

$$p(u + v) \leq p(u) + p(v) \qquad \text{(subadditivty)}$$

and

$$p(\alpha v) = |\alpha| p(v) \qquad \text{(absolute homogeneity)}$$

for all $u, v \in V$ and $\alpha \in \mathbb{K}$.

Definition C.2. We say that two seminorms p_1 and p_2 on a vector space are *equivalent* if there exist two constants $c, C > 0$ such that

$$c p_1 \leq p_2 \leq C p_1.$$

Remark C.3. It is clear that for a given vector space this definition induces an equivalence relation on the set of all seminorms on that vector space.

The following lemma is well-known, see, e. g., [35, Exercise 36, p.206].

Lemma C.4. *Let p_1 and p_2 two seminorms on a finite-dimensional vector space. Then p_1 and p_2 are equivalent if and only if $\ker(p_1) = \ker(p_2)$.*

Proof. If p_1 and p_2 are equivalent then it is clear that $\ker(p_1) = \ker(p_2)$. Let $\ker(p_1) = \ker(p_2)$ and call the domain of p_1 and p_2 the vector space V. Then p_1 and p_2 are norms on the quotient space $V/\ker(p_1)$. Since all norms on a finite-dimensional vector space are equivalent the norms p_1 and p_2 on $V/\ker(p_1)$ are equivalent. This implies that also the seminorms p_1 and p_2 on V are equivalent. $\qquad \square$

D. Miscellaneous results

In [10, p. 440] the Kronecker product is defined.

Definition D.1 (Kronecker product). Let $A = (a_{ij}) \in \mathbb{C}^{m \times n}$ and $B = (b_{ij}) \in \mathbb{C}^{p \times q}$. Then, the *Kronecker product* $A \otimes B \in \mathbb{C}^{(mp) \times (nq)}$ of A and B is the partitioned matrix

$$A \otimes B := \begin{bmatrix} a_{11}B & \cdots & a_{1n}B \\ \vdots & \ddots & \vdots \\ a_{m1}B & \cdots & a_{mn}B \end{bmatrix}.$$

Remark D.2. If we say $v \in \mathbb{C}^n$, then v is a column vector, i. e. $v \in \mathbb{C}^{n \times 1}$. Thus, the Kronecker product $A \otimes B$ is also defined if $A \in \mathbb{C}^n$ or $B \in \mathbb{C}^n$. For the basic properties of the Kronecker product we refer to [10].

Lemma D.3. *For all* $m, n \in \mathbb{N}$ *let* $P_{m,n} \in O(mn)$ *be the Kronecker permutation matrix such that*

$$P_{p,m}(A \otimes B)P_{n,q} = B \otimes A \qquad \text{for all } A \in \mathbb{C}^{m \times n} \text{ and } B \in \mathbb{C}^{p \times q},$$

see [10, Fact 7.4.30]. For all natural numbers $m, n_1, \ldots, n_k \in \mathbb{N}$ *let* $Q_{m,n_1,\ldots,n_k} \in O(m(n_1 + \cdots + n_k))$ *be the permutation matrix* $(P_{m,n_1} \oplus \cdots \oplus P_{m,n_k})P_{n_1+\cdots+n_k,m}$. *Then the following statements hold:*

(i) For all $A_i \in \mathbb{C}^{m_i \times n_i}$, $i \in \{1, \ldots, k\}$, *and* $B \in \mathbb{C}^{p \times q}$ *we have*

$$(A_1 \oplus \cdots \oplus A_k) \otimes B = (A_1 \otimes B) \oplus \cdots \oplus (A_k \otimes B).$$

(ii) For all $A \in \mathbb{C}^{m \times n}$ *and* $B_i \in \mathbb{C}^{p_i \times q_i}$, $i \in \{1, \ldots k\}$, *we have*

$$A \otimes (B_1 \oplus \cdots \oplus B_k) = Q_{m,p_1,\ldots,p_k}^{\mathsf{T}}((A \otimes B_1) \oplus \cdots \oplus (A \otimes B_k))Q_{n,q_1,\ldots,q_k}.$$

(iii) For all $A \in \mathbb{C}^{m \times n}$ *and* $B_1, \ldots, B_k \in \mathbb{C}^{p \times q}$ *we have*

$$A \otimes (B_1 \oplus \cdots \oplus B_k) = (P_{m,k} \otimes I_p)((A \otimes B_1) \oplus \cdots \oplus (A \otimes B_k))(P_{k,n} \otimes I_q).$$

Proof. (i) This is easy to check.

(ii) For all $A \in \mathbb{C}^{m \times n}$ and $B_i \in \mathbb{C}^{p_i \times q_i}$, $i \in \{1, \ldots, k\}$, we have

$$
\begin{aligned}
A \otimes (B_1 &\oplus \cdots \oplus B_k) \\
&= P_{m, p_1 + \cdots + p_k}((B_1 \oplus \cdots \oplus B_k) \otimes A) P_{q_1 + \cdots + q_k, n} \\
&= P_{m, p_1 + \cdots + p_k}((B_1 \otimes A) \oplus \cdots \oplus (B_k \otimes A)) P_{q_1 + \cdots + q_k, n} \\
&= P_{m, p_1 + \cdots + p_k}((P_{p_1, m}(A \otimes B_1) P_{n, q_1}) \oplus \cdots \\
&\qquad \oplus (P_{p_k, m}(A \otimes B_k) P_{n, q_k})) P_{q_1 + \cdots + q_k, n} \\
&= Q^{\mathsf{T}}_{m, p_1, \ldots, p_k}((A \otimes B_1) \oplus \cdots \oplus (A \otimes B_k)) Q_{n, q_1, \ldots, q_k}.
\end{aligned}
$$

(iii) By Fact 7.4.30 *viii*) in [10] we have

$$
Q_{n, q, \ldots, q} = (I_k \otimes P_{n, q}) P_{kq, n} = P_{k, n} \otimes I_q. \qquad \square
$$

It is well-known that commuting orthogonal matrices are simultaneously quasidiagonalisable:

Theorem D.4. *Let $\mathcal{S} \subset O(n)$ be a nonempty commuting family of real orthogonal matrices. Then there exist a real orthogonal matrix Q and a nonnegative integer q such that, for each $A \in \mathcal{S}$, $Q^{\mathsf{T}} A Q$ is a real quasidiagonal matrix of the form*

$$
\Lambda(A) \oplus R(\theta_1(A)) \oplus \cdots \oplus R(\theta_q(A))
$$

in which each $\Lambda(A) = \mathrm{diag}(\pm 1, \ldots, \pm 1) \in \mathbb{R}^{(n-2q) \times (n-2q)}$, $R(\theta)$ is the rotation matrix $\left(\begin{smallmatrix} \cos \theta & -\sin \theta \\ \sin \theta & \cos \theta \end{smallmatrix} \right)$ and each $\theta_j(A) \in [0, 2\pi)$.

Proof. This follows immediately by [39, Corollary 2.5.11.(c), Theorem 2.5.15]. $\qquad \square$

We now state Kronecker's approximation theorem, see, e. g., Corollary 2 on page 20 in [38].

Theorem D.5 (Kronecker's approximation theorem). *For each irrational number α the set of numbers $\{\alpha n \text{ reduced modulo } 1 \,|\, n \in \mathbb{N}\}$ is dense in the whole interval $[0, 1)$.*

We also need Turán's third theorem, see Theorem 11.1 on page 126 in [57].

Theorem D.6 (Turán's third theorem). *Let $b_1, \ldots, b_n, z_1, \ldots, z_n \in \mathbb{C}$. If m is a nonnegative integer and the z_j are restricted by*

$$
\frac{\min_{\mu \neq \nu} |z_\mu - z_\nu|}{\max_j |z_j|} \geq \delta \, (> 0), \quad z_j \neq 0
$$

then the inequality

$$\max_{\nu=m+1,\ldots,m+n} \frac{\left|\sum_{j=1}^{n} b_j z_j^{\nu}\right|}{\sum_{j=1}^{n} |b_j|\,|z_j|^{\nu}} \geq \frac{1}{n}\left(\frac{\delta}{2}\right)^{n-1}$$

holds.

We also need Theorem A.1 of [4].

Theorem D.7. *Let (X, d) be a metric space, $(Y, \mathcal{F}, \mu)$ be a measure space and $f \colon X \times Y \to \mathbb{R}$ be such that*

(i) $f(x, \,\cdot\,)$ *is μ-integrable for all $x \in X$,*

(ii) $f(\,\cdot\,, y)$ *is continuous in X for μ-almost all $y \in Y$,*

(iii) *there exists $m \in L^1(Y, \mu)$ satisfying*

$$\sup_{x \in X} |f(x, y)| \leq m(y) \qquad \text{for } \mu\text{-a.e. } y \in Y.$$

Then the map

$$X \to \mathbb{R}, \quad x \mapsto \int_Y f(x, y)\,d\mu(y)$$

is bounded and continuous.

Bibliography

[1] A. Aghaei, K. Dayal, and R. S. Elliott. Symmetry-adapted phonon analysis of nanotubes. *Journal of the Mechanics and Physics of Solids*, 61(2):557–578, 2013.

[2] R. Alicandro and M. Cicalese. A general integral representation result for continuum limits of discrete energies with superlinear growth. *SIAM J. Math. Anal.*, 36(1):1–37, 2004.

[3] J. L. Alperin and R. B. Bell. *Groups and representations*, volume 162 of *Graduate Texts in Mathematics*. Springer-Verlag, New York, 1995.

[4] L. Ambrosio, G. Da Prato, and A. Mennucci. *Introduction to measure theory and integration*, volume 10 of *Appunti. Scuola Normale Superiore di Pisa (Nuova Serie)*. Edizioni della Normale, Pisa, 2011.

[5] M. I. Aroyo, editor. *International tables for crystallography. Vol. A. Space-group symmetry*. Hoboken, NJ: John Wiley & Sons, sixth revised edition, 2016.

[6] M. Arroyo and T. Belytschko. An atomistic-based finite deformation membrane for single layer crystalline films. *J. Mech. Phys. Solids*, 50(9):1941–1977, 2002.

[7] H. Bacry, L. Michel, and J. Zak. Symmetry and classification of energy bands in crystals. In *Group theoretical methods in physics (Varna, 1987)*, volume 313 of *Lecture Notes in Phys.*, pages 291–308. Springer, Berlin, 1988.

[8] Z. Bai, J. Demmel, J. Dongarra, A. Ruhe, and H. van der Vorst, editors. *Templates for the solution of algebraic eigenvalue problems*, volume 11 of *Software, Environments, and Tools*. Society for Industrial and Applied Mathematics (SIAM), Philadelphia, PA, 2000.

[9] E. Behrends. *Introduction to Markov chains*. Advanced Lectures in Mathematics. Friedr. Vieweg & Sohn, Braunschweig, 2000.

[10] D. S. Bernstein. *Matrix mathematics.* Princeton University Press, Princeton, NJ, second edition, 2009.

[11] J. L. Birman. *Theory of crystal space groups and lattice dynamics.* Springer-Verlag, Berlin, 1984.

[12] X. Blanc, C. Le Bris, and P.-L. Lions. Atomistic to continuum limits for computational materials science. *M2AN Math. Model. Numer. Anal.*, 41(2):391–426, 2007.

[13] C. J. Bradley and A. P. Cracknell. *The mathematical theory of symmetry in solids.* Oxford Classic Texts in the Physical Sciences. The Clarendon Press, Oxford University Press, New York, 2010.

[14] A. Braides and M. S. Gelli. Continuum limits of discrete systems without convexity hypotheses. *Math. Mech. Solids*, 7(1):41–66, 2002.

[15] J. Braun. Connecting atomistic and continuous models of elastodynamics. *Arch. Ration. Mech. Anal.*, 224(3):907–953, 2017.

[16] J. Braun and B. Schmidt. On the passage from atomistic systems to nonlinear elasticity theory for general multi-body potentials with p-growth. *Netw. Heterog. Media*, 8(4):879–912, 2013.

[17] J. Braun and B. Schmidt. Existence and convergence of solutions of the boundary value problem in atomistic and continuum nonlinear elasticity theory. *Calc. Var. Partial Differential Equations*, 55(5):Art. 125, 36, 2016.

[18] H. Brown, R. Bülow, J. Neubüser, H. Wondratschek, and H. Zassenhaus. *Crystallographic groups of four-dimensional space.* Wiley-Interscience, New York-Chichester-Brisbane, 1978.

[19] L. S. Charlap. *Bieberbach groups and flat manifolds.* Universitext. Springer-Verlag, New York, 1986.

[20] A. Clark. *Elements of Abstract Algebra.* Dover Books on Mathematics Series. Dover Publications, 1984.

[21] S. Conti, G. Dolzmann, B. Kirchheim, and S. Müller. Sufficient conditions for the validity of the Cauchy-Born rule close to $SO(n)$. *J. Eur. Math. Soc. (JEMS)*, 8(3):515–530, 2006.

[22] M. Dresselhaus, G. Dresselhaus, and R. Saito. Physics of carbon nanotubes. *Carbon*, 33(7):883–891, 1995.

[23] W. E and P. Ming. Cauchy-Born rule and the stability of crystalline solids: dynamic problems. *Acta Math. Appl. Sin. Engl. Ser.*, 23(4):529–550, 2007.

[24] W. E and P. Ming. Cauchy-Born rule and the stability of crystalline solids: static problems. *Arch. Ration. Mech. Anal.*, 183(2):241–297, 2007.

[25] D. El Kass and R. Monneau. Atomic to continuum passage for nanotubes: a discrete Saint-Venant principle and error estimates. *Arch. Ration. Mech. Anal.*, 213(1):25–128, 2014.

[26] R. S. Elliott, N. Triantafyllidis, and J. A. Shaw. Stability of crystalline solids. I. Continuum and atomic lattice considerations. *J. Mech. Phys. Solids*, 54(1):161–192, 2006.

[27] J. L. Ericksen. On the Cauchy-Born rule. *Math. Mech. Solids*, 13(3-4):199–220, 2008.

[28] J. M. G. Fell and R. S. Doran. *Representations of *-algebras, locally compact groups, and Banach *-algebraic bundles. Vol. 2*, volume 126 of *Pure and Applied Mathematics*. Academic Press, Inc., Boston, MA, 1988.

[29] W. Florek, D. Lipiński, and T. Lulek, editors. *Symmetry and Structural Properties of Condensed Matter: Second International School of Theoretical Physics; Poznań, Poland, 26 August - 2 September 1992*. World Scientific, 1993.

[30] M. Friedrich, E. Mainini, P. Piovano, and U. Stefanelli. Characterization of optimal carbon nanotubes under stretching and validation of the Cauchy-Born rule. *Arch. Ration. Mech. Anal.*, 231(1):465–517, 2019.

[31] M. Friedrich and U. Stefanelli. Graphene ground states. *Z. Angew. Math. Phys.*, 69(3):Art. 70, 18, 2018.

[32] G. Friesecke, R. D. James, and S. Müller. A hierarchy of plate models derived from nonlinear elasticity by gamma-convergence. *Arch. Ration. Mech. Anal.*, 180(2):183–236, 2006.

[33] G. Friesecke and F. Theil. Validity and failure of the Cauchy-Born hypothesis in a two-dimensional mass-spring lattice. *J. Nonlinear Sci.*, 12(5):445–478, 2002.

[34] F. R. Gantmacher. *The theory of matrices. Vol. 1.* AMS Chelsea Publishing, Providence, RI, 1998.

[35] I. M. Glazman and J. I. Ljubič. *Finite-dimensional linear analysis.* Dover Publications, Inc., Mineola, NY, 2006.

[36] L. C. Grove and C. T. Benson. *Finite reflection groups*, volume 99 of *Graduate Texts in Mathematics.* Springer-Verlag, New York, second edition, 1985.

[37] C. Heil. *A basis theory primer.* Applied and Numerical Harmonic Analysis. Birkhäuser/Springer, New York, expanded edition, 2011.

[38] E. Hlawka, J. Schoissengeier, and R. Taschner. *Geometric and analytic number theory.* Universitext. Springer-Verlag, Berlin, 1991.

[39] R. A. Horn and C. R. Johnson. *Matrix analysis.* Cambridge University Press, Cambridge, second edition, 2013.

[40] T. Hudson and C. Ortner. On the stability of Bravais lattices and their Cauchy-Born approximations. *ESAIM Math. Model. Numer. Anal.*, 46(1):81–110, 2012.

[41] R. D. James. Objective structures. *J. Mech. Phys. Solids*, 54(11):2354–2390, 2006.

[42] D. Jüstel. *Radiation for the Analysis of Molecular Structures with Non-Crystalline Symmetry: Modelling and Representation Theoretic Design.* Dissertation, Technische Universität München, München, 2014.

[43] E. Kaniuth and K. F. Taylor. *Induced representations of locally compact groups*, volume 197 of *Cambridge Tracts in Mathematics.* Cambridge University Press, Cambridge, 2013.

[44] J. Martinet. *Perfect lattices in Euclidean spaces*, volume 327 of *Grundlehren der Mathematischen Wissenschaften.* Springer-Verlag, Berlin, 2003.

[45] C. C. Moore. Groups with finite dimensional irreducible representations. *Trans. Amer. Math. Soc.*, 166:401–410, 1972.

[46] D. Olson and C. Ortner. Regularity and locality of point defects in multilattices. *Appl. Math. Res. Express. AMRX*, 2017(2):297–337, 2017.

[47] W. Opechowski. *Crystallographic and metacrystallographic groups.* North-Holland Publishing Co., Amsterdam, 1986.

[48] C. Ortner and F. Theil. Justification of the Cauchy-Born approximation of elastodynamics. *Arch. Ration. Mech. Anal.*, 207(3):1025–1073, 2013.

[49] J. G. Ratcliffe. *Foundations of hyperbolic manifolds*, volume 149 of *Graduate Texts in Mathematics.* Springer, New York, second edition, 2006.

[50] F. Riesz and B. Sz.-Nagy. *Functional analysis.* Frederick Ungar Publishing Co., New York, 1955.

[51] B. Schmidt. A derivation of continuum nonlinear plate theory from atomistic models. *Multiscale Model. Simul.*, 5(2):664–694, 2006.

[52] B. Schmidt. On the passage from atomic to continuum theory for thin films. *Arch. Ration. Mech. Anal.*, 190(1):1–55, 2008.

[53] B. Schmidt. On the derivation of linear elasticity from atomistic models. *Netw. Heterog. Media*, 4(4):789–812, 2009.

[54] B. Simon. *Representations of finite and compact groups*, volume 10 of *Graduate Studies in Mathematics.* American Mathematical Society, Providence, RI, 1996.

[55] B. Steinberg. *Representation theory of finite groups.* Universitext. Springer, New York, 2012.

[56] H. Streitwolf. *Group theory in solid-state physics.* University physics series. Macdonald and Co., 1971.

[57] P. Turán. *On a new method of analysis and its applications.* Pure and Applied Mathematics (New York). John Wiley & Sons, Inc., New York, 1984.

[58] D. S. Watkins. *The matrix eigenvalue problem.* Society for Industrial and Applied Mathematics (SIAM), Philadelphia, PA, 2007.

[59] R. Webster. *Convexity.* Oxford Science Publications. The Clarendon Press, Oxford University Press, New York, 1994.

[60] J. Yang and W. E. Generalized cauchy-born rules for elastic deformation of sheets, plates, and rods: Derivation of continuum models from atomistic models. *Phys. Rev. B*, 74, 11 2006.

In der Reihe *Augsburger Schriften zur Mathematik, Physik und Informatik,* herausgegeben von Prof. Dr. B. Aulbach, Prof. Dr. F. Pukelsheim, Prof. Dr. W. Reif, Prof. Dr. B. Schmidt, Prof. Dr. D. Vollhardt,

sind bisher erschienen:

18 Olga Birkmeier — Machtindizes und Fairness-Kriterien in gewichteten Abstimmungssystemen mit Enthaltungen

ISBN 978-3-8325-2968-0, 2011, 153 S.   37.50 €

19 Johannes Neher — A posteriori error estimation for hybridized mixed and discontinuous Galerkin methods

ISBN 978-3-8325-3088-4, 2012, 105 S.   33.50 €

20 Andreas Krug — Extension groups of tautological sheaves on Hilbert schemes of points on surfaces

ISBN 978-3-8325-3254-3, 2012, 127 S.   34.00 €

21 Isabella Graf — Multiscale modeling and homogenization of reaction-diffusion systems involving biological surfaces

ISBN 978-3-8325-3397-7, 2013, 285 S.   46.50 €

22 Franz Vogler — Derived Manifolds from Functors of Points

ISBN 978-3-8325-3405-9, 2013, 158 S.   35.00 €

23 Kai-Friederike Oelbermann — Biproportionale Divisormethoden und der Algorithmus der alternierenden Skalierung

ISBN 978-3-8325-3456-1, 2013, 91 S.   32.50 €

24 Markus Göhl — Der durchschnittliche Rechenaufwand des Simplexverfahrens unter einem verallgemeinerten Rotationssymmetriemodell

ISBN 978-3-8325-3531-5, 2013, 141 S.   34.50 €

25 Fabian Reffel — Konvergenzverhalten des iterativen proportionalen Anpassungsverfahrens im Fall kontinuierlicher Maße und im Fall diskreter Maße

ISBN 978-3-8325-3652-7, 2014, 185 S.   40.00 €

26 Emanuel Schnalzger — Lineare Optimierung mit dem Schatteneckenalgorithmus im Kontext probabilistischer Analysen

ISBN 978-3-8325-3788-3, 2014, 225 S.   42.00 €

35	Johanna Kerler-Back	Dynamic iteration and model order reduction for magneto-quasistatic systems
		ISBN 978-3-8325-4910-7, 2019, 175 S. 35.50 €
36	Veronika Antonie Auer-Volkmann	Eigendamage: An Eigendeformation Model for the Variational Approximation of Cohesive Fracture
		ISBN 978-3-8325-4969-5, 2019, 151 S. 39.00 €
37	Miguel de Benito Delgado	Effective two dimensional theories for multi-layered plates
		ISBN 978-3-8325-4984-8, 2019, 153 S. 38.00 €
38	Martin Steinbach	On the Stability of Objective Structures
		ISBN 978-3-8325-5378-4, 2021, 161 S. 40.00 €

Alle erschienenen Bücher können unter der angegebenen ISBN im Buchhandel oder direkt beim Logos Verlag Berlin (www.logos-verlag.de, Fax: 030 - 42 85 10 92) bestellt werden.